BUSINESS
— AND —
ENTREPRENEUR
SUCCESS
MANUAL

David H. Dudley, PE

HOWARD PUBLISHING

BUSINESS AND ENTREPRENEUR SUCCESS MANUAL

ISBN: 978-0-9998304-0-6 (Hardback, color version)
ISBN: 978-0-9998304-4-4 (Paperback)
ISBN: 978-0-9998304-5-1 (ePUB, color version)
Also avilable for Kindle.

Published by:
Howard Publishing

CONTENTS

DEDICATION

This book is dedicated to every person, young and old, employed and unemployed, educated and those with little schooling, who dream of becoming better off financially, and aspire to build a happy, successful, and rewarding life. It doesn't matter whether you live in the city, suburbs, or country; this book was written with the hope and intention of enabling you to bring your dreams to fruition.

ACKNOWLEDGMENTS

I owe a tremendous amount of gratitude to my family for their love, and encouragement, patience, and sacrifices they have made in my life.

I am also indebted to the University of Oklahoma, the University of Tennessee, and Sacramento State University, as well as the faculty of these fine institutions, for their valuable contributions in the way of information, knowledge, and guidance. Thanks also to all the graduate students who assisted me along the way by conducting an extensive amount of research and providing critical data for this book.

I would like to thank my previous employers, peers, and co-workers with whom I have learned much. Their mentorship and trust in allowing me the opportunity to work on and manage all types of projects has provided me with real world experience and equipped me with the skill set needed to better understand best business principles and share the information with others in a practical way.

I would also like to thank all of those who have taught me so much about best business principles, management, marketing, entrepreneurship, and business over the past five decades. I beg forgiveness for not mentioning you individually as the list is too long and, quite frankly, I have simply forgotten some names. Nevertheless, your time, wisdom, instruction, and influence in my life is appreciated more than words can ever say.

In addition, I must thank many of the associations, organizations, and alternative media sources that are dedicated to providing truth. Because of their efforts, we are able to get the real story without a government political agenda and mainstream media bias. Further acknowledgement, and other resource information, about these indispensable groups is provided in the Appendix.

I greatly appreciate Mitzi Houser for her professional assistance in proofreading this book for grammatical errors. Global Deziners must also be acknowledged and complemented for the excellent work they did in creating the book cover.

Lastly, my deepest gratitude and greatest acknowledgement is reserved to the one and only God of everything, in and through my Lord and Savior, Jesus Christ. Only through His grace, provisions, forgiveness, and work in my life is anything I do possible. I fully recognize and appreciate that God is in control. I am at awe, and undeserving, of the unconditional love, mercy, grace, and immeasurable blessings He has bestowed upon me. All praise and thanks for everything (including this book) must go to Him.

INTRODUCTION

Success - the correct or desired result of something attempted

Have you ever wondered why some people achieve success while others endlessly search for it? Everyone strives for success. Why do some achieve it and others don't? What's the secret?

> **"If you are going to achieve excellence in big things, you develop the habit in little matters."** – Colin Powell, an American elder statesman and a retired four-star general in the United States Army. Powell was born in Harlem, NY

The qualities needed to be successful in business are often the same ones we utilize to be successful in our personal lives. Examples of these include time management, integrity, and leadership. These are crucial to business success. The chapters that follow will show you how to develop and utilize these characteristics in the work place.

> **"If you look to lead, invest at least 40% of your time managing yourself – your ethics, character, principles, purpose, motivation, and conduct."** – Dee Hock, founder and former CEO of Visa credit card association

Many skills such as prioritization, organization, and persuasion are skills we use in our day-to-day lives and are also key to success in the workplace. This book will give you tips and insights on how to apply these and others in the workplace to ensure business success.

> **"It's the details that are vital. Little things make big things happen."** – John Wooden, won 10 NCAA national championships in a 12-year period as head coach at UCLA

While many traits and skills such as those mentioned above are known and used by most people, there are also business skills that may not be as familiar; yet, they are instrumental in success. A few of these include networking, marketing, and branding. These will be defined and discussed in depth in the following pages.

"It takes 20 years to build a reputation and five minutes to ruin it. If you think about that, you'll do things differently."
– Warren Buffett, chairman and CEO Berkshire Hathaway

When you have the right tools to develop these characteristics and skills as a person and utilize them in the workplace, you will be successful in both your personal and professional life. This book is designed to fill your tool box with the tools essential for success.

"The secret of getting ahead is getting started." – Mark Twain, an American writer, humorist, entrepreneur, publisher, and lecturer.

PART I

Cultivating Success

CHAPTER 1

Integrity

What is Integrity?

Standard definitions of integrity include "the quality of being honest and having strong moral principles" and "a firm adherence to a code of especially moral or artistic values." In simpler terms integrity means doing the right thing, because it's the right thing to do; telling the truth, even if it isn't pretty or isn't convenient for you; and taking responsibility for your words and actions.

Integrity is NOT situational. It is not okay to compromise your integrity or go against your beliefs on "little things" thinking they don't matter. Once you start making compromises in small decisions, it becomes easier to do so in larger situations, as the line becomes blurred.

It is doing the right thing under all circumstances—in a crowd-filled room or an empty office. Integrity should be the building block for all businesses.

Why Should Integrity be Important to You and Your Business?

Having honesty and integrity in the workplace is one of the most important qualities of great leadership in business. The first value that all the successful executives agree on is integrity. Leaders know that honesty and integrity are the foundations of leadership. Leaders stand up for what they believe in.

There are many examples of temporary winners who won by cheating. For example, Enron was cited as one of America's most innovating and daring companies. The CEO of the company knew and worked with some of the most important people in the country, including the President of the United States. Unfortunately, Enron's success was built on lies. As always, the consequence from the lack of integrity always catches up. Enron declared bankruptcy. Thousands of people were thrown out of work, and thousands of investors—including most of the company's employees—lost billions of dollars as Enron's shares shrank to penny-stock levels.

Leaders with integrity may not be the most famous or flashy of leaders, and they don't care. Integrity means doing the right thing because it is the right thing to do. Leaders keep their promises. They give promises carefully, even reluctantly, but once they have given that promise, they follow through on that promise without fail. And they always tell the truth. Those ingredients add up to success.

As President Eisenhower once said, "The supreme quality for leadership is unquestionably integrity.

Without it, no real success is possible..." Therefore, integrity is of the utmost importance in both our personal and professional lives, because we would all like to be successful in both.

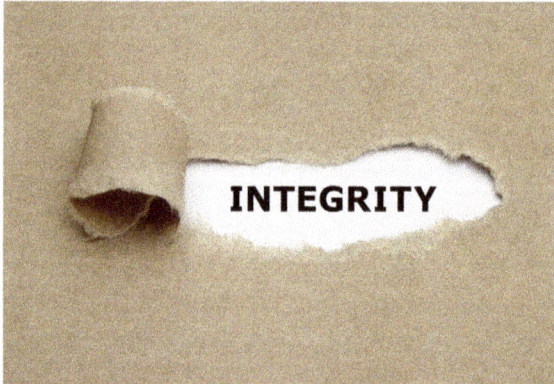

Integrity in a business starts at the top and trickles down. If the person starting a company does not act with integrity, others involved in the company will typically follow suit, and the company will fail. It can take years to build a positive, trustworthy reputation both personally and professionally, but it only takes a second or one act to destroy it.

Acting with integrity will lead to the following:

1. Positive reputation
2. Trust
3. Better relationships
4. Respect
5. Loyalty
6. Increased Profit Potential

In the bigger picture, acting with integrity will help you do your part to improve the world. Being a person or business known for acting with integrity will bring you all of the above, but more importantly, acting with integrity raises the bar for those around you. As an individual you make the people around you expect more from the people around them, and as a business you force other businesses in your industry to be mindful of their actions or risk losing their customers, partners, and investors.

Roadblocks to Integrity

Most everyone knows integrity is important both personally and professionally. In fact most businesses have mission statements that include the word integrity or something akin to it. So, why doesn't every person and individual act with integrity?

There are three main roadblocks to acting with integrity:

1. Rationalization—humans have the innate ability to explain away just about anything. Let's face it—people try to rationalize murder, so they will definitely rationalize other decisions in their lives and businesses. In all honesty, this is a tough one. Do you tell potential clients/customers EVERYTHING about your product? Are you completely honest with your boss/supervisor when asked for feedback? These are just a sample of the choices we are faced with every day dealing with integrity, and we can rationalize whichever way we choose to act upon them.

2. Easy way out—it is much easier to not talk about your mistakes, even when confronted with them. Many people will blame someone/something else. You wouldn't have been late for the appointment or missed the meeting, if your phone hadn't rung. No one told you there was a meeting, even if it was posted on the calendar. Your co-worker handled the part of the product that malfunctioned.

3. Want it now—in today's society, people want instant gratification. They don't want to have to work for it for months or even years. Therefore, they are tempted to take shortcuts that diminish the integrity of their product, business, and person.

Interestingly enough, while the norms of today's society encourage the roadblocks to acting with integrity, the technological advancements of today's society illuminate and punish those who do not act with integrity. Personal and business lives are transparent, whether we want them to be or not. As soon as you interact with someone on a professional or personal level, the feeling with which the other person walked away from that interaction will be on the internet for everyone in the world to see, especially if it was a bad feeling.

How to Achieve Integrity

Fortunately, integrity is not something we're born with or without. It is a behavior learned over time, and we can set goals to ensure we act with more integrity every day. The following are characteristics to develop to ensure you are acting with integrity:

1. **Keep promises**—do what you say you will do.
2. **Be honest**—always tell the truth, even when it hurts.
3. **Be able to handle honesty**—you need to be able to handle honesty when shown to you; see the situation as it really is, not the way you want it to be
4. **Know that you could be wrong—this is especially** true for supervisors. Supervisors need to act with confidence, but they also should be aware of the possibility that they could be wrong.
5. **Treat others with respect**—take time to get to know people.
6. **Build and maintain trust**—relationships (both personal and professional) are based on trust.
7. **Don't take shortcuts**—deliver the best product/service possible.
8. **Build a solid reputation**—this comes from developing everything else on this list and takes time.
9. **Praise others for accomplishments**—be genuinely happy and proud of others.
10. **Own up to mistakes**—apologize when you make a mistake, are late for a meeting, etc. When possible correct your mistakes immediately.
11. **Be consistent**—people can count on you to be the same today, tomorrow, and a month from now.
12. **No blame game**—don't blame other people when things go wrong.
13. **Don't gossip**—mind your own business; take care of your own responsibilities; don't worry about what other people are doing.
14. **Be skeptical of simple solutions**—don't accept a simple solution for a complex problem without thoroughly examining it; don't partake in fads or get caught up in buzzwords.
15. **Be able to forgive**—realize that no one is perfect, and everyone (including you) needs forgiveness at times.

When your business embodies the characteristics listed above, it will have a good, solid reputation; loyal customers; ethical employees; potential for making more money; and, it will force other businesses to act with more integrity.

In Conclusion

Remember, when building your personal or professional integrity, the people and businesses with whom you associate are often viewed as mirrors of you, and others will judge you by their actions.

Building integrity takes time, but that integrity can be destroyed with one act in one second. This is not to say you won't make mistakes. Part of integrity is how you deal with mistakes, when they are made. As Alan K. Simpson (interestingly enough—a former politician) said, "If you have integrity, nothing else matters. If you don't have integrity, nothing else matters."

CHAPTER 2

Positive Thinking for Success

Every single thought you allow in your mind has a ripple effect not only on the way you act but on the things you invite into your life. This principle applies to family, love, health, happiness and even work.

A positive attitude in business helps one accomplish tasks faster and in a better manner. The owner is better able to manage the company and inspire the staff, when he is not bogged down by the challenges.

The performance of employees, to a great extent, depends on their attitudes. There is a direct relationship between attitude and production. The better the attitude the more production. Higher production also occurs when employees demonstrate a positive attitude towards their superiors, their work, and their colleagues. Through positive energy work becomes a pleasure, and employees find it easier to achieve their goals.

A positive attitude has significant benefits for an individual in many aspects. Some of these benefits are listed below.

Benefits of a Positive Attitude at Work

1. **Less stress**. To begin with, it's healthier for you and everyone around you. Stress has a detrimental effect on the health of employees. More and more studies are showing that stress can bring about a host of physical and mental problems from insomnia, fatigue, and loss of concentration to more serious ailments like severe depression, bodily aches and pains, hypertension (high blood pressure), digestive disorders (as severe as ulcers), and even heart attacks and strokes. You need to nip it in the bud now to avoid both short- and long-term side effects. Stress can be reduced through positive thinking, and with reduced stress, employees will enjoy better health and take fewer sick leaves.

2. **Career success:** Employees' success in the workplace is measured through their performance. Employees with a positive attitude will always think of ways to accomplish their task in a well-defined manner instead of complaining or finding excuses for non-performance. This results

in success either through promotion or increased compensation.

3. **Productivity:** With a positive attitude, employees tend to take more interest in what they do and deliver. Consequently, they produce better quality work with minimum errors. This improves their overall output as well as productivity.

4. **Leadership:** Working in an organization is all about managing a diverse workforce. Some employees earn respect easily, and people often follow and listen to them. This is possible through the positive attitude demonstrated by leaders.

5. **Team work:** Good relationships among employees help build effective teams in which all members are united and work for a common cause. A positive attitude helps employees appreciate each other's competencies and work as a team to achieve common objectives instead of being overly perturbed by inadequacies of other team members.

6. **Decision making:** Having a positive attitude helps employees make better decisions in an objective manner. It triggers a healthy thought process, enabling employees to choose wisely and logically.

7. **Motivation:** Having a positive attitude helps motivate employees to overcome obstacles they may face during the course of their job. It also determines the way they see the world around them. The moment they are successful in overcoming obstacles, they are motivated to move forward.

8. **Interpersonal relations:** Customers prefer to deal with someone who is positive in nature. A positive attitude enables employees to develop a better rapport with customers, earning valuable customer loyalty.

In conclusion, a positive attitude at work is beneficial not only to the organization, but also to the owner and employees on an individual basis.

If you take responsibility for yourself, your actions, your work, your attitude every single day, and you make it the best that you can, you are going to overcome roadblocks and be rewarded with much greater success.

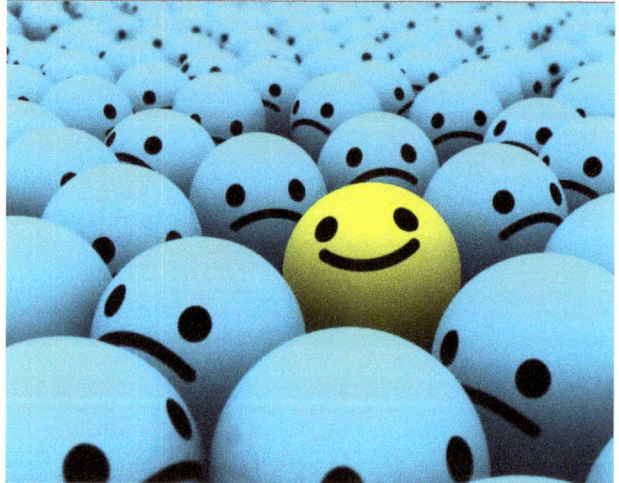

What is the impact of your attitude on your business?

Your attitude rubs off on existing and potential customers, your staff, your suppliers, your investors and all those with whom you come into contact.

Maintaining a positive attitude is infectious; those around you will pick up on your positive energy. Everyone in your company will feel positive, and customers will want to do business with you. This, in turn, will lead to you maximizing the performance of your business.

If you maintain a negative attitude, the opposite is likely to happen. People will not want to be around you; your staff will feel demotivated; and customers will not want to buy from you. The result will be deteriorating performance of your business.

A positive approach will give you control and confidence, helping you perform at your best, whereas a negative approach will damage confidence, harm performance, paralyze your mental skills and may also impact your health.

How to maintain a positive attitude?

Look for the positive side. Although it is easy to focus on the negative with so much negative news around you must look for the positive and focus on this.

Keep an open mind. Always be open to the possibility that a life change you have refused to consider might be the key to transforming your life for the better.

This type of attitude impresses your colleagues. Why? Because most of them have been faced with the same challenge and chose to not change. Their attitude towards the change has been clouded with self-doubt and lack of courage.

When you are willing to keep an open mind, you are responding positively to the challenge of a life change with the possibility of a great reward.

Be aware of your thoughts. Focus on how you think about different situations. If you see a list of businesses closing down in your sector, do you think your business is also doomed, or do you see it as an opportunity to find new customers and maybe even increase market shares? Challenge any negative thoughts and turn them into positive ones.

Find your inner coach. What you think about yourself, your abilities, your financial gains (or losses) and your assessment of your ability to succeed becomes what you could actively live out. Actions follow thoughts.

To be successful, choose your thoughts wisely and deliberately. When your attitude is unproductive,and you need to shift from a negative thought pattern, draw upon your inner coach: Transform fearful, defeatist thinking into thoughts charged with power, conviction and faith. Focus on your ability to meet your goals.

Learn to coach yourself into a productive state of mind. This coaching can lead to forward movement and be reignited anytime you face a setback.

Think of your attitude with this equation: Energy = Motion. From this paradigm, attitude determines movement forward or back.

Present a positive attitude to others. Even if you are feeling down, make sure you present a positive attitude to everyone with whom you interact. This will help you feel more positive and make others feel positive, too.

Focus on what has gone well. Focus on what has gone well and what you have achieved. View things that have not gone so well as learning experiences.

See the glass
~HALF FULL~
Then fill it
the rest of
the way

Determine Incentives. Discover what motivates you to take action. Think back to what has inspired you to make changes in your life and pursue your goals. Knowing why you do what you do is vital to your current motivation and passion.

Is your incentive financial gain? Is it self-preservation, stability, anger or fear? Having clear incentives can instantly shift your attitude to a positive one.

When you can motivate yourself toward a goal, your attitude becomes infused with your inner power, enthusiasm and passion. Your outlook on business is elevated. You walk faster, smile more and carry a posture of self-assuredness, all of which draws even more success your way.

Soar with the Eagles. Surround yourself with people who have a positive attitude and can help you be the best you can be. Seek out those who can help you be more successful.

Ignore whiners and complainers. Whiners and complainers see the world through crap-colored glasses. They'd rather talk about what's irreparably wrong, rather than make things better. More importantly, complainers can't bear to see anyone else happy and satisfied.

Create a "library" of positive thoughts. Spend at least 15 minutes every morning reading, viewing, or listening to something inspirational or motivational. If you do this regularly, you'll have those thoughts and feelings ready at hand (or rather, ready to mind) when events don't go exactly the way you'd prefer.

Focus on improving the performance of your business. Focus on getting your business in the best shape to weather the downturns and to maximize your short-term and long-term business performance.

Be proactive, not reactive. A reactive individual is at the mercy of change. A proactive individual sees change as a part of the process and takes action to make the best of it.

Having a proactive attitude requires work. You must be able to think ahead and anticipate. It involves being involved.

In business (and life) you cannot simply sit back and let things just happen as they will. In truth, you could, but that attitude is a negative response that influences negative action, namely, reaction.

Think big. If you think small, you will achieve small. If you think big, you are more likely to achieve a goal beyond your wildest dreams.

When we allow ourselves to have an attitude that pushes boundaries and explores possibilities, we draw in people who have the same attitude. In other words, by thinking big we find big thinkers.

Want to have a team full of big thinkers? Want to have meetings where ideas are shared and positive plans are made? Want to grow leaders out of your team and promote them to new heights in their career? It all starts with your big-thinking, boundary-pushing, dream-inspiring attitude.

Keep your business goals at the forefront of your mind. What are the goals for your business? Revisit them and make sure they are still relevant. Make sure all your current actions support your goals.

Enter Action with Boldness. When you do something, do it boldly and with confidence so you make your mark. Wimping out is more likely to leave you stuck in the same old pattern -immune to positive change.

In the end it's all about getting things done—with a positive attitude. As leaders, we need to be able to move and work with a certain sense of boldness. A boldness that inspires us and those around us to reach for new horizons in all we do.

It's obvious, action is better than no action—but bold action that leaves a mark is what we should be doing in our life and business.

Set Benchmarks. Select benchmarks to help you visualize goals as right before you as you strive to develop your skills. To maintain an attitude of enthusiasm, create benchmarks of success and think about them intentionally and consistently.

The more deeply you believe a goal will come to pass, the more likely it will. This type of mental practice keeps your attitude positive and directed toward the attainment of success. I believe you naturally strive toward what you visualize.

Manage Anxiety. When challenges surface in business, you might experience a loss of control over emotions and thoughts. To remain optimistic, discipline your mind to stay clear of catastrophic thoughts and the "what if" slippery slope of anxiety.

Anxiety is simply fear projected forward. Remind yourself that a "what if" is not happening now: It is just a story. Focus on what you can control right now. Each thought is a seed you plant that programs your attitude and your subsequent behavior. To keep a great attitude, learn to focus your mind on solutions, not problems.

Create a Positive Impression. To reach your peak potential on the climb to success, be driven and shrewd but also happy and enjoyable to work for and with. As the heights of attained success grow, the more infectious your energy appears to other people.

Your personal vibe can either repel people or draw them in.

Carrying an attitude of excitement and vigor transfers to those around you and increases business morale. If your attitude is gregarious, infectious and all about living life to its fullest, you appear to have unlimited potential: You positively motivate others to support your attempt to achieve even higher levels of success.

Be Hungry. Be hungry for success. This hunger will make a natural eagerness permeate your attitude. It will seem as though you cannot wait to get to work and put that hunger into motion.

This hunger becomes the emotional "speaker of the house." This energy serves as the driving force in attaining your goals.

Give thanks in all circumstances. Being thankful in all circumstances helps us be overcomers. Being thankful turns weaknesses and hardships into strengths, opportunities, and successes.

Keep a gratitude journal. For some, one single event can ruin an entire day, or an unpleasant interaction or experience at night can overshadow the enjoyable parts of the day. Instead of clinging to the negative, it is important to intentionally focus on the good parts of the day to offset the negative. Try writing down things for which you are grateful every day and see how your attitude changes. Science has found that gratitude can significantly increase our happiness, and protect us from stress, negativity, anxiety and depression.

Use positive words to describe your life. The words we use have more power than we think. How we think and talk about our life is how our life will be. If we describe our life as boring, busy, mundane, chaotic, that is how we will perceive it, and we will feel the effects in our body and mind. If we use the words simple, involved, familiar or lively, we will see our life in a whole different light and find more enjoyment.

Don't let yourself get dragged into other people's complaints. Your day was going pretty well until you got to work and your co-worker wouldn't stop complaining about this or that. Don't let anyone pull you into their negativity. A study done at the Warsaw School of Social Psychology shows that complaining leads to lower moods and negative emotions, decreased life satisfaction and optimism, and emotional and motivational deficits.

Make someone else smile. Who do you think about most of the time? If we answered honestly, most of us would say ourselves. It's good to hold ourselves accountable, take responsibility for our life roles, hygiene, food, etc., but we should set a goal each day to make someone else smile. We need to think about someone else's happiness, and it will help us realize the immense impact our attitude and expression has on the people around us.

NO NEGATIVITY ALLOWED HERE

Key Attitudes

There are five key attitudes t small businesses should emulate for maximum success. A company should also try to hire employees with these attitudes and train employees to strengthen these attitudes. A company with a culture of these attitudes will prosper far above the competition that does not embrace these attitudes as standard operating philosophy.

1. **Respectfulness**—Respect is a very important attitude in the workplace and doesn't solely pertain to employee interaction with management. Employees should also have a respectful attitude when interacting with clients, customers and co-workers. Employees with this type of attitude are willing to treat other people politely and professionally, even if they disagree with the other person's point of view.

2. **Commitment**—A committed attitude is a valued one in the workplace. Small businesses need employees who are not only committed to the goals and initiatives that affect the bottom line, but who also are committed to their particular positions. Employees project a committed attitude by showing a willingness to do whatever it takes to fulfill the duties of their positions and, via the development of new ideas, to make the company even better.

3. **Innovation**—Employees with an innovative attitude don't shy away from trying something new or finding a different way to do things. Small businesses need employees who think outside of the box and develop new ways to accomplish tasks and approach goals. Employees with this type of attitude know their ideas might not work out to be the best way to do something, but that the biggest failure is in not at least giving new ideas a shot.

4. **Helpfulness**—It is important to have a helpful attitude at work, whether that means assisting clients and customers with their needs or helping co-workers accomplish overall company goals. The more helpful an attitude employees have, the more people want to be around them at work and the more willing they are to partner with those employees on key projects and initiatives.

5. **Be Appreciative**—It is always nicer to get a pat on the back than to get stabbed in the back. If you're wondering "how do I change my negative attitude," start by observing how you communicate with coworkers. If you practice asking useful questions, giving accolades and being gracious at work for two weeks, you'll notice a difference in the people you work with and in your own feelings about work.

As a Man Thinketh

- Men do not attract what they want, but what they are.
- A man is literally what he thinks; his character is the sum of all his thoughts.
- Cherish your visions. Cherish your ideals. Cherish the music that stirs in your heart, the beauty that forms in your mind, the loveliness that drapes your purest thoughts; for out of them will grow all delightful conditions, all heavenly environment;

of these, if you but remain true to them, your world will at last be built.

- The soul attracts that which it secretly harbors, that which it loves, and also that which it fears. It reaches the height of its cherished aspirations; it falls to the level of its unchastened desires; and circumstances are the means by which the soul receives its own.
- Men are anxious to improve their circumstances but are unwilling to improve themselves; they, therefore, remain bound.
- Every action and feeling is preceded by a thought.
- Right thinking begins with the words we say to ourselves.
- Circumstance does not make the man; it reveals him to himself.
- You cannot travel within and stand still without.
- As the physically weak man can make himself strong by careful and patient training, so the man of weak thoughts, can make himself and his life better by exercising right thinking.
- Every man is where he is by the law of his being;. The thoughts which he has built into his character have brought him there, and in the arrangement of his life there is no element of chance, but all is the result of a law which cannot err.
- The thoughtless, the ignorant, and indolent, seeing only the apparent effects of things and not the things themselves, talk of law, of fortune, and chance. Seeing a man grow rich, they say, 'How lucky he is!' Observing another become intellectual they exclaim, 'How highly favored he is!' And noting the saintly character and wide influence of another, they remark, 'How chance aids him at every turn!' They don't see the trials and failures and the struggles which these men have voluntarily encountered in order to gain their experience. They have no knowledge of the sacrifices these men have made, of the undaunted efforts they have put forth, of the faith they have exercised, that they might overcome the apparently insurmountable and realize the vision of their heart. They do not know the darkness and the heart aches; they only see the light and the Joy, and they call it 'luck;' they do not see the long arduous journey, but only behold the pleasant goal and call it 'good fortune;' they do not understand the process, but only perceive the result, and call it 'chance.'
- The circumstances which a man encounters with suffering are the result of his own mental harmony.

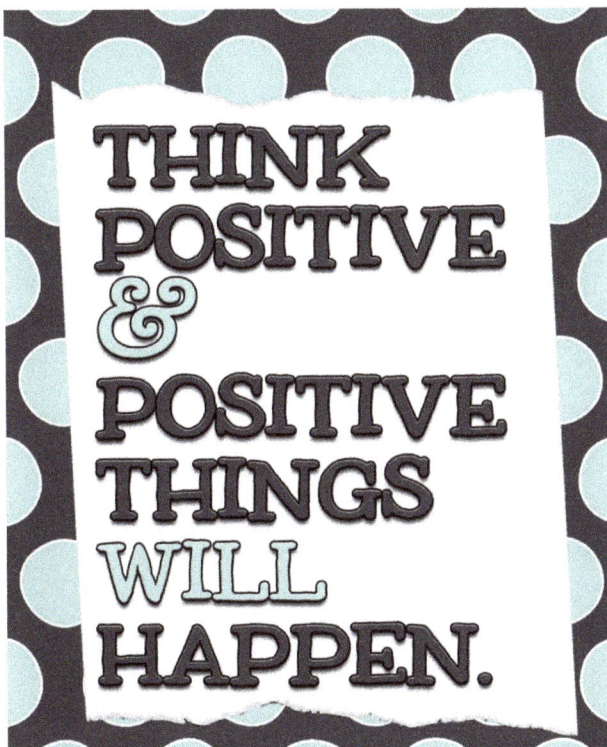

Remember that YOU Control Your Thoughts and Attitude

Attitude does not emerge from what happens to you but instead from how you decide to interpret what happens to you. Develop and make every effort to maintain a consistently positive attitude for better enjoyment in life, stronger relationships, and optimal business success.

CHAPTER 3

Success: What Successful People Have in Common & How to Be Successful

Success

What is success? Success has many different definitions, depending on who you ask. The dictionary defines success as the accomplishment of one's goal; the desired result of an attempt; or one that succeeds. Some people claim success is achieved when you are making a positive contribution to other people or causes. Pop culture would have us believe that success is all about money, fame, and power. The numerous stories of unfulfilled lives among so called "successful people" shows, however, that success is not solely about achievements or possessions.

According to Earl Nightingale's definition, a success is anyone who is realizing a worthy predetermined ideal. So, a success is the farmer who is growing crops, because that's what he wants to do. A success is the entrepreneur who starts her own company, because that was her dream.

Happiness

Research in psychology has shown there are six essential aspects of **"happiness"** in life:

1. **Physical Health**: We need to be physically healthy to have the energy to engage in life.

2. **Mental Fitness**: We need to be continuously engaging our minds: learning, growing, experiencing new ideas, getting better, pursuing mastery, and putting our ideas to work to accomplish our goals.

3. **Emotional Health**: We need to be self-aware emotionally, feel good about ourselves, and have a positive self-image.

4. **Social Health**: We need positive relationships in our life along with people who love and support us. Humans are designed to be social creatures. Having friends and loved ones who support and encourage us to be a better person and whom we can trust helps us achieve happiness.

5. **Purpose/Meaning/Spiritual Health:** We tend to be happier when we make a positive impact on others' lives;it gives meaning and purpose to our work and daily life.

6. **Material Wealth**: There is a basic level of food, shelter, and clothing that all people need and is paid for with money. If we are too poor or have too much stress from struggling financially, we may find it more difficult to be happy.

We can function and be happy in the short term without these elements, but in the long term if we are missing or have a short supply of any of them, our ability to be truly happy people will most likely be negatively impacted. If we are unhappy, it is our body's way of telling us something is wrong with our life, and we should do something about it.

Success and happiness, happiness and success. People have a tendency to confuse the two, mixing them up until they are unable to discern the very different but seemingly unified terms. Following are a list of key points to help distinguish the differences between success and happiness.

1. **Success is meeting deadlines. Happiness is working toward your goals.**
2. **Success is working your way to the top of your field. Happiness is flowing in your purpose and gifts.** When you figure out for what you've been created and start doing it, you will find a satisfaction and contentment you never imagined possible.
3. **Success is focusing on accumulating wealth. Happiness is focusing on improving your life.**
4. **Success is promotion above your peers. Happiness is being respected by your peers.** Being promoted is a good thing to work towards, but clawing your way ahead, no matter who you have to step on or use to get there, will not bring you lasting satisfaction or peace.
5. **Success is to live driven. Happiness is to live according to your passion.** The most successful and happy people are those who identify what they love to do and figure out how to do it.

The Key to Being Successful and Happy

Having a personal definition of what success means to you, and setting goals accordingly, is the only way to make your achievements truly meaningful which coincides with the ability to be happy. Striving to succeed according to somebody else's standard or

definition may win you the admiration of others, but will it bring you a sense of personal accomplishment and happiness?

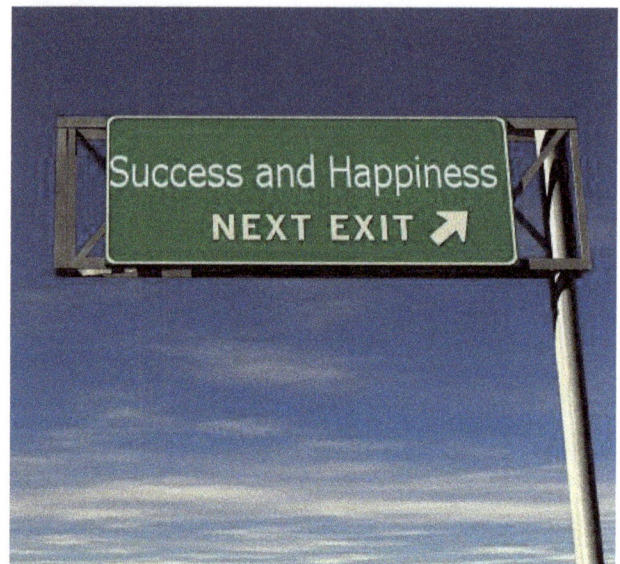

If you want to be a success on a level that makes you feel truly satisfied, then take some time right now to write down your personal definition of success. Be specific for each area of your life.

> **"Success is not the key to happiness. Happiness is the key to success. If you love what you are doing, you will be successful."** -Albert Schweitzer

Think about what being financially and emotionally successful would feel like. What kind of relationships would represent successful relationships to you? Once you've achieved some success in these areas, how will you spend your time, your money, and your energy? Who would you spend them with?

Having a clearly defined picture of where you want to be in all areas of life will provide a definite purpose and make evident the most efficient path to both happiness and success.

Author's Note: Please keep in mind that Happiness and Success are worldly terms, and should not be confused with contentment. Many people try to fill a void or a deep pain in their lives, and in doing so, unfortunately try to fill that void with things that can't satisfy. Whenever we look to fill a void with anything other than God we only end up needing and wanting it more. Whether it be relationships, sex, substances, money, status, popularity, fame, career, etc. it only leaves one feeling even more empty and depressed than when they started.

Many things that are used in an attempt to fill a void or block out pain aren't necessarily bad things in themselves, some can even be positive, or good for others, the community, and/or the environment. However, when they become the end goals, good or bad, and the reason for our being, they result in discontentment, because those things were never meant to fulfill us.

The only place that we can really find true fulfillment and contentment is living a life in harmony with God, loving Him above all, in and through Christ. True contentment isn't something that we find in things, people or circumstances, it can only be found in God, who bestows upon us immeasurable blessings, in whom gave us life and sustains us, who is our helper and comforter, who provides wisdom, peace, mercy, forgiveness, grace, and unconditional love.

True contentment is having an understanding that God's power, purpose, plan, and provisions are sufficient for every circumstance. Being able to live through all kinds of adversity, believing in and experiencing God's sufficiency, through Christ, brings true contentment and peace. We have to choose to trust in God and His plan despite what may be going on in our lives. Then, and only then, will true contentment be achieved

What Successful People Have in Common & How to Be Successful

Successful people talk, think, and approach situations, challenges, and problems differently than most people, and the only way to be successful is to take the same actions successful people take. Success is no different than any other skill. Duplicate the actions and mindsets of successful people, and you will create success for yourself.

The publications *Entrepreneur, Success, Forbes,* and *MoneyTalks*, as well as the U.S. Small Business Administration and the Brian Tracy organization have examined, in depth, the characteristics prominent business leaders and successful entrepreneurs have in common. They have identified characteristics shared by these successful people.

Following are the most commonly found qualities, personality traits, and habits of successful people. As you read this list, honestly consider how each characteristic or trait pertains to you. Prepare a plan for improving on the areas you feel could use improvement.

1. **Big dreams** – It is rare to read a biography of a successful person who didn't have grand dreams of what they wanted to accomplish. Sir Richard Branson, Walt Disney and Sam Walton all had big dreams, and, for the most part, achieved even more than they originally imagined. This is one of the reasons they became successful. They were not afraid to dream big and then go for it. If you want to know whether or not someone will be successful, ask them about their dreams. If you want to be successful, be sure to dream big.

2. **Goal-oriented.** Successful people don't wait for life to happen to them. They have a plan on how they want their life to go, and then they set

goals to make that plan a reality. Henry Ford said, "Nothing is particularly hard if you divide it into small jobs." Breaking down big goals into bite-size pieces can pave the way for success.

3. **Plan everything.** Planning every aspect of life and business is necessary for success. The act of business planning is importan,t because it requires you to analyze each business situation, research and compile data, and make conclusions based mainly on facts revealed through research. A plan for business, and life, serves as a guide to help you efficiently reach your goals. A plan is a map which takes you from point A to Z, a means to help you measure your progress, a motivator, and a spirit lifter as you see progression. Plans should be made for every aspect of life, for every day, week, month, year and five years indicating what you need to do to accomplish your goals. Tomorrow's "to do" list is a plan to make a day's progress towards your goal. What you need to accomplish this week, next month, and so on to accomplish your goals should be written down and posted where you will see them. Successful people make plans that help them reach their goals.

4. **Persistent.** Successful people don't quit when faced with difficulties or obstacles. Instead, they keep plugging along, and it is that persistence that helps them achieve success.

5. **Inquisitive.** Successful people are lifelong learners. They recognize there is always more to learn and that understanding gives them the tools and resources needed to reach their goals.

6. **Self-aware.** People who find success are the ones who know who they are. They know their abilities as well as their limitations. As a result, they focus on where they excel, and they find the right people with different talents for the tasks outside their expertise.

7. **Risk-takers.** Successful people are risk-takers;. however, the risks they take are calculated ones.

When they fail, successful people take the time to learn from their mistakes.

8. **Quick-thinking and decisive.** Slow and steady may have won the race for the turtle, but it's not a particularly good strategy when it comes to human success.

Much has been written detailing how the model of business success is moving from one of stability to transient competitive advantages. In fact, an Ernst & Young survey revealed that 81 percent of consumer products executives stated fast decision-making was of critical importance to business success.

However, being decisive is important outside the business world as well. If you stall for years while trying to decide whether to move your family, change careers or go back to school, you lose precious time that could have been used to build your success. Make your decisions promptly and move forward without looking back.

9. **Good communicators.** Successful people are also good communicators. They articulate their ideas clearly. If you are concerned you are not an effective communicator, take heart. Good communication is a learned skill that anyone can obtain.

10. **Enthusiastic.** Enthusiastic people create success, because they are passionate. They love what they do.

Also, people naturally gravitate toward those who radiate excitement and self-confidence. As a result, rather than working alone, an enthusiastic person garners a tribe of people working toward the same goal.

11. **Self-disciplined.** The most successful people among us are those who can keep themselves focused and on task. They are not easily deterred or distracted from the things that help them reach their goals.

12. **Hard-working.** Successful people work hard. A 2016 U.S. Trust study surveyed individuals

with investable assets of at least $3 million. It found that 77 percent of wealthy individuals reported growing up in middle-class families or poorer. That means most financially successfully people didn't have their money handed to them — instead, they likely worked hard for it.

The best things in life — whether money in the bank or a great relationship with your spouse or child — takes significant effort.

13. **Positive attitude.** Successful people have realistic optimism—realistic because they take action and optimism because they believe their success is inevitable. This positive attitude allows them to persevere and be resilient when things don't go their way.

14. **Flexible**. One misconception ordinary people have about persevering is that to persevere you must stay the course no matter what. This is true only if the reason for pursuing your goal is still valid. Many successful people became successful doing something different from their original intentions. The world is always changing, and successful people know when to be flexible and change course.

15. **Masters of time**. Successful people are successful, because they get a lot done. The only way to do that is to get the most out of each day. Successful people value their time and see the direct connection between how they spend their time, how it affects their well-being and how it leads to the accomplishment of their goals. They are cognizant of time and manage time wisely.

They prepare and stick to a schedule whenever possible. They are able to say "no" in situations contrary to their goals. They are good at adhering to strict start and end times for meetings.

16. **Action oriented**. Those who don't achieve much with their lives tend to use "would," "should" and "could" repeatedly. Additionally, we all know people who have tremendous potential and talk a big game but accomplish very little. Successful people are doers. They don't wait for conditions to be perfect before they take action. They go for it and modify along the way if necessary; but, they keep doing everything possible to reach their goals. Those who get what they want are the ones who take action to reach their goals.

17. **Confident**. Successful people are confident. Confidence helps you achieve your goals, which in turn makes you more confident. A great way to be confident is remembering your past successes. Signs of confidence include empowering others, not being offended by criticism, and understanding that the first time you do something is always the hardest. Success is the combination of confidence and competence.

18. **Readers**. Most, if not all, successful people like to read. If it is your desire to become a successful person, reading should be a part of your daily life. With access to more books than could be read in several lifetimes, it is important to read books that will make the most difference to you. Usually these are the books that enable you become a true expert in your passion, or help you grow and improve as a person.

19. **Trusts intuition.** People who succeed in life trust their "gut." They may not be able to explain rationally why or how they made their decision, but they knew it was the right thing to do at the time. Successful people learn to harness the power of their subconscience. This involves mentally visualizing the outcome beforehand. Meditating, rather than looking at our smart phones, is a great

way to think about challenges and get in touch with our intuition.

20. **Well-rounded and balanced**. Successful people strive to be successful in all aspects of their lives. They tend to live healthy lives, while becoming financially independent, nurturing meaningful relationships, and developing personal mastery while working towards their professional goals. They know sacrificing one key area to achieve another will not help them maximize their true potential.

21. **Excellent network.** Successful people understand the importance of relationships and realize that relationships with the right people play a very important factor in achieving many goals. They work hard to develop and nurture relationships with the right people.

22. **Admit mistakes and take responsibility**. Successful people admit when they are wrong allowing them to focus on the solution. Blaming others and making excuses are two things that will almost guarantee you will NOT be successful in the future. When we blame others or make excuses, we give up our responsibility and power to make improvement.

23. **Good character**. A person with good character exhibits attributes such as integrity, honesty, courage, loyalty, fortitude, and other virtues that promote good behavior and habits. Most successful people tend to have good character.

24. **Great company**. Successful people surround themselves with the right kind of people. Studies have shown that the people with whom we spend time, play an incredibly important role in our levels of success, health, and happiness.

Some people can be parasites. They drain your valuable time and suck out your happiness, energy, and possibly some of your tangible resources as well. So what makes someone a "good" person to spend time with? And what are the benefits of surrounding yourself with these people? They are people that will inspire you to be a better person, provide you with motivation to achieve your goals, empower you to make the changes you need to succeed and cheer on your success.

In the workplace good people tend to be productive people. They're organized, create schedules and don't get easily distracted from the end goal. Learning the habits of these types of people can help you be more productive.

It's important to note that "good" does not necessarily mean they are saints or similar to you. Hanging out with people similar to you can inhibit growth. We need to have an eagerness to soak up knowledge and experiencing diverse perspectives can help us do that. However, it doesn't mean we need to learn or emulate every aspect of that person's life. It simply means finding out what the person does to be exceptional in a particular area can help us improve that aspect of our own life.

Surrounding ourselves with people from whom we can learn and grow facilitates success. The more we surround ourselves with those that inspire and educate us in matters that coincide with our goals, the faster we will achieve success. Surrounding ourselves with people living the life we want to live helps us learn their habits and styles. It's wonderful to have co-workers, friends, family and loved ones who support and motivate you, but there's nothing like acquiring the knowledge from those who make you a better person.

25. **Listen.** Successful people are good at listening to others. Everyone wants to be a great speaker, but few strive to be excellent listeners. People who sincerely inquire and actively listen succeed in life, because they amass knowledge and are able to understand the real needs of other people. Furthermore, the fastest way to be a good conversationalist is to ask questions and listen well.

26. **Self-control**. Successful people rarely lose control. They don't go into panics or blind rages. They have learned to control their emotions and to consciously (or subconsciously) put themselves in a resourceful state. They understand they can't change other people or what's happening to them, but they can change how they react and how they feel about it. Another sign of self control is doing what you're supposed to do despite how you feel about it.

27. **Prepared**. Successful people have a knack for being prepared. They mentally rehearse and visualize the possibilities vividly so that when the actual situation occurs, their brains will "remember" what to do.

28. **Choice**. People who are successful are in control. They know they always have a choice. They don't feel victimized by their circumstances, and they truly believe the past does not determine the future. They write the script to their lives.

29. **Self-reliant**. Successful people believe in themselves and their ability to achieve their dreams whether anyone helps them or not. They also don't allow other people to slow them down.

30. **Energy conscious**. Those who become great understand that successfully managing their energy is just as important as managing their time. One of the most important principles in energy management is knowing that rest is as important as action. Successful people are aware that low energy produces poor results, and this violates their need for excellence. As much as possible they also work on the most critical tasks when their energy is highest.

31. **Manage money wisely.** People who are financially successful are actively involved with their money, and they understand what their assets are doing. They have budgets, they track spending, they ensure they regularly contribute to investments, and they plan ahead to avoid financial pitfalls.

32. **Ask for the sale.** In a way, we are all selling something, whether it be a product, a service, a viewpoint, or just ourselves in a favorable way. To be successful, you have to ask for the sale. And the most successful people are good at not letting opportunities get away. Successful people know how to close.

33. **Focus on the customer.** Successful people and businesses realize customers drive the success or failure of the enterprise. They understand and focus on delivering excellent customer service. They are also good at identifying new and creative ways to maintain and expand customer bases in this exceedingly competitive marketplace.

    For most industries, it is a researched fact that most business (approximately 80 percent) will come from repeat customers rather than new customers. Therefore, along with trying to draw newcomers, the more you can do to woo your regular customers, the better off you will be in the long run. Personalized attention and an expression of genuine gratitude is very much appreciated and remembered in a world where these characteristics seem to be disappearing at an alarming rate.

34. **Self-promoters.** One of the greatest myths about personal or business success is that eventually your business, personal abilities, products or services will get discovered and be embraced by the masses who will beat a path to your door to buy what you are selling. But how can this happen, if no one knows who you are, what you sell or why they should be buying?

    Hard work is the foundation of success, but it's not enough. If you're going to get the opportunities you want, you need to make those stellar results visible to others.

    Self-promotion is one of the most beneficial, yet most underutilized, marketing tools business leaders and entrepreneurs have at their immediate disposal.

Many negative connotations typically come to mind when thinking about self-promotion, all of which unfortunately keep most people from feeling comfortable and confident when talking about themselves. We all know the immature favorites: conceited, show-off, braggart, arrogant, egotistical. Nobody wants to fall victim to such name-calling.

Yet, if you're not comfortable claiming your achievements and promoting yourself, it will be difficult, if not impossible, to get ahead. Most of us were told growing up that if we worked hard and did a good job (at school, in athletics, or elsewhere), we'd be recognized for our effort and rewarded accordingly. For the most part, it was good advice. However, in today's competitive workplace, if your plan to get ahead is based on the assumption that hard work alone will suffice, you may find yourself being left behind as the horn blowers around you land the opportunities you anticipated were yours.

Successful people are tactful self-promoters. Research shows t people who are comfortable promoting themselves are not only more successful in landing a job from an interview, but also in building stronger networks—within their organization and outside of it—and they are more successful in business and throughout their careers. That's not to say that humility is no longer a virtue. But false humility can leave you languishing. Self-promotion is about strategically building your "personal brand" or "personal objective" to ensure those who can help you reach your goals will know who you are, the value you have, and/ or the opportunity you offer in front of them.

35. **Projecting a positive image.** You have but a passing moment to make a positive and memorable impression on people. Cultivating a positive image is a key factor for success. It involves self-awareness, knowledge of how to look, how to dress in the range of settings, how to behave, and how to interact at all levels. It is rare to find a successful person who is not good at projecting a positive image.

36. **Level the playing field with technology.** Many people have become successful by learning how to use technology to their advantage. Though previously constrained by small budgets and workforces, technology can level the playing field for small businesses. Today's technology provides advanced reporting and business analysis capability so one can gain a deep understanding of business performance, customer preferences, and market trends. It is also a great way to efficiently communicate messages to one or many people.

37. **Build a top-notch team.** No one person can be successful, or build a successful business, alone. It is a task that requires a team –people who can provide assistance along the way. A team may include family members, friends, suppliers, business alliances, employees, sub-contractors, industry and business associations, local government and the community. Team members can also be customers or clients.

For a team to be most effective, it needs a good leader. The better the team, the more success the leader will likely experience. Picking the right people for your team, and leading them effectively, will result in the highest probability of success.

38. **Known as an expert.** When you have a problem that needs to be solved, do you seek just anyone's advice, or do you seek an expert in the field to help solve your particular problem? Obviously, you want the most accurate information and assistance you can get. You naturally seek an expert to help solve your problem. Therefore, it only stands to reason that the more you become known for your expertise in your business, the more people will seek you out to tap into your expertise; which in turn opens up opportunities.

39. **Create a competitive advantage.** Successful people and businesses have often created a competitive advantage through one or several means. It could be outstanding customer service, a longer warranty, better selection, longer business hours, more flexible payment options, lowest price, personalized service, a network, an outstanding team, user-friendly/better return and exchange policies, or a combination of several of these elements that allows for the greatest success.

40. **Personal investment.** Successful people read books, magazines, reports, journals, newsletters, websites and industry publications that help them improve and be more successful. They join business associations and clubs and develop networks with the right people to learn more and to further facilitate their goals and objectives. They also attend seminars, workshops and training courses to better equip themselves with the knowledge and skills needed to be more successful. They do this, because they know education is an ongoing process. They understand there are usually ways to do things better, in less time, with less effort. In short, people who are successful invest in themselves.

41. **Accessibility.** To get ahead one must be accessible to the right people at the right time. Accessibility in meeting with people to discuss opportunities, mentor you or do business with you is critical for success. However, if you are going to maintain a margin of your time for your most important priorities, you have to make some tough decisions regarding your accessibility. Successful people master their time and accessibility to align with their priorities in order to achieve their goals as efficiently as possible.

42. **Good reputation.** A good reputation is a tremendous asset most successful people have in common. You can't simply buy a good reputation; it's something you earn by honoring your promises, holding to an honorable level of integrity, and treating people fairly.

A person or company's good reputation is essential for success. Think about it, who wants to do business with a person or company with a poor reputation? If you build up and keep a good reputation people will want to do business with you.

43. **Involved.** Most successful people get involved in the community, associations, clubs, and other outside endeavors that coincide with their business and/or personal goals. Truly, successful people know that doing it all alone can only take you so far. Our human brains are wired for connection, and we need the support of others to persevere and endure the disappointments on the road to success. We can also learn much from the examples and advice of people who have travelled the path before us, as well as from the mistakes and victories of colleagues travelling the same path presently.

44. **Good at negotiations.** The ability to negotiate effectively is unquestionably a skill everyone who desires to be successful in business must make every effort to master. Successful people recognize that negotiating is a process, not just something that is done when discussing the terms and conditions of a solution. Negotiating is much more than haggling about price. It requires an

understanding of the dynamics that affect the process and influence the behavior of people. Successful people look for ways to solve problems and grow opportunities. They also know how and when to limit their concessions so they can work out an agreement equitable for both parties.

Too many people search for the quick fix to close the sale or resolve the problem. Successful people recognize that patience is a virtue and rushing the process often leads to an undesirable outcome. They don't hurry to reach an agreement. Instead, they take time to gather the necessary information. They think carefully about possible solutions.

45. **Conducive workspace.** Successful people create workspaces that help them be successful. A properly setup workspace has an empowering effect and facilitates productivity.

Most people spend forty hours per week at their workspace. Therefore, it is important the place be an inspiring, pleasant environment, suitable for maximum productivity. A study by the American Society of Interior Designers showed a direct correlation between those who like their workspace and job satisfaction. The survey also found that 38% of workers cited a comfortable chair as one of the most important elements of a pleasant working environment. In addition, a workspace setup to reduce distractions—such as social media—will provide the best results.

Another key element to consider is lighting. Good lighting is one of the most important factors contributing to a pleasant working environment. Furthermore, large windows that let in a good deal of natural daylight are beneficial and preferred by most people.

Cluttered doesn't function as well as clean. Those with cluttered and messy desks waste an extraordinary amount of time looking for things. Stay organized so you can enjoy your space.

One of the best things you can do to decrease your stress load, inspire your creative juices, and optimize productivity is to put together a workspace that is conducive to helping you reach your goals. If you need further convincing of the importance of having a dedicated workspace, consider the financial benefits and/or the value of achieving your goals sooner.

46. **Get and stay organized.** Highly successful people stand apart from the rest. They rise to the top as others take notice that they are getting the job done. Highly successful people are more effective than most people. They engage in good habits that lead to success including staying organized. Organization may be important to regular workers, but to the effective, highly successful person, being organized is vital to goal attainment.

47. **Limit the number of hats.** The phrase "wearing too many hats" simply means being involved in too many different roles or having too many tasks to perform (taking on too much). Successful people often became successful after realizing they couldn't do everything, focusing on priorities, and building a team of people who could help them reach their goals.

It is easy to get caught up in taking on too much and quickly becoming overloaded. When we overload our life, we become less effective.

The perception may be it is cheaper to do everything yourself or that others won't do it exactly how you want it done. This view may be true, but this approach fails to take into consideration all the things not getting done, missed opportunities, and costly postponements of goal attainment (i.e. costly in time and revenue). Also, the right person for the role may bring in new ideas and do it even better than imagined.

Avoid the trap of taking on too much and wearing too many hats. This approach will help you see the "big" picture better, work better, and reach your goals faster. Remember that if you say "yes" to something, you have to say "no" to something else. If you are already overloaded you have to stop adding more opportunities and volunteering. Focus on priorities and seek out qualified assistance.

48. **Follow-up constantly.** Successful people and successful businesses follow-up and follow-through with customers, prospects, business alliances, and opportunities. Constant and consistent follow-up enables you to turn prospects into customers and bring opportunities into fruition. This approach will increase sales, increase the value of each sale, increase the frequency from existing customers, and build stronger business relationships with suppliers and your core business team. Follow-up is especially important with your existing customer base, as the real work begins after the sale.

49. **Demonstrate courage.** Observance of successful people reveals that they carry themselves with an air of confidence and conviction. Courage is not the absence of fear, but the willingness to confront your fear and do something about it. Fear is normal. Fear is common. Fear is human. However, fear is a success killer. We're all afraid at times, but successful people face their fears and act.

50. **Embrace change.** The successful look at how the world is shifting and apply this to how they may grow their advantage. They know they must continue to adapt. This willingness to accept change is a great quality of the successful.

51. **Have a high level of motivation.** Motivation refers to the act or state of being stimulated toward action. Highly successful people continually seek and uncover reasons to stay perpetually provoked to new levels of success. This motivation often comes from a passion or love of the endeavor. Other times it is driven by a vision of goal attainment (the perspective of how things will be after the goal is accomplished).

To succeed, it is critical that you be stimulated, excited, and driven to some action or actions. This isn't the kind of enthusiasm that lasts for a few hours, a day, or a week; it is based on what you do each day to stimulate yourself toward actions and inspire yourself to keep going. Motivation can be drawn from many sources including a positive attitude, peer pressure, goals, passion, responsibility, fear, and reward. Cultivate healthy attributes to get and stay motivated.

52. **Interested in results.** Successful people rarely place value on effort, work, or time spent on an activity; they value the results.

53. **Highly ethical.** Success and ethics can co-exist and must co-exist for a person to be truly successful. It is when these two do not co-exist that problems occur. Every choice has a positive reward or a negative consequence. The result of choice may not happen immediately, but it will surface sooner or later. It may be as simple as having peace of mind or mental stress. It could be something like a new welcome opportunity or a financial setback. The result of ethical choices will come about one way or another, and it is much more enjoyable to obtain the positive rewards of doing the right thing.

54. **Don't worry about judgment from others.** Successful people don't need approval from others to feel happy, validated, or confident. They don't worry about what others have to say—good or bad - because they know how to find confidence and security internally, not externally.

55. **Care about others.** Successful people care about others and the larger community. A common trait among most of those who attain top positions or build great organizations is a willingness to help others achieve their goals. They teach, listen and develop people. When so busy with our own *stuff*, it is easy to forget the needs of others, and the impact we have on them. Truly successful people and great leaders don't forget about others.

It is important to take the time to listen and learn about others, and we should always live by the golden rule: "Do unto others as you would have them do unto you." Also, helping others reach their dreams can help us reach our dreams. And remember the truism, *'People don't care how much you know until they know how much you care'.*

56. **Rather fail than not try.** The old saying, "no risk, no reward" seems to apply to those who are successful. Successful people go for it despite the risk of failure. They know that if they don't put themselves in a position to fail, they'll never create the ability to win.

Unsuccessful people seem to play it safe. They may not speak up or offer ideas, because they operate from a place of fear. They may be afraid to fai,l because they're overly concerned with the judgment of others and do the minimum to "fly under the radar." Never be afraid of failure, because behind every mistake is an opportunity to learn, grow, develop, and be successful.

57. **"Reach up" in relationships.** Successful people don't hesitate to reach out and up to other people. They realize these connections and the value others bring to their lives opens up more opportunities and facilitates goal achievement.

Make a habit of "reaching up" in all of your relationships—reach up toward people who are better connected, better educated, and even

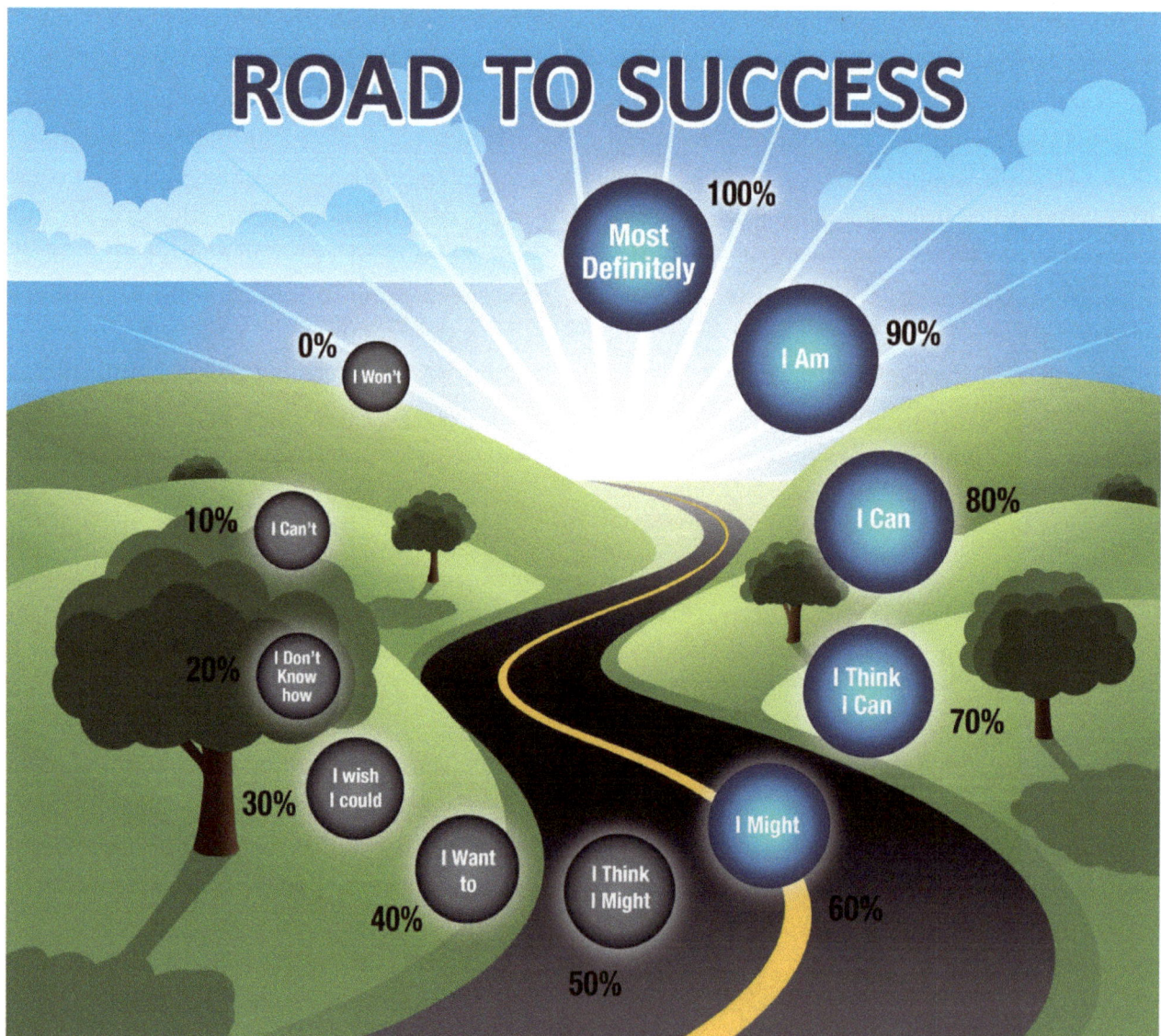

ROAD TO SUCCESS

- 100% Most Definitely
- 90% I Am
- 80% I Can
- 70% I Think I Can
- 60% I Might
- 50% I Think I Might
- 40% I Want to
- 30% I wish I could
- 20% I Don't Know how
- 10% I Can't
- 0% I Won't

more successful. The people with whom you surround yourself will have a great deal to do with whether you achieve your goals—or not. Going up, by associating with bigger thinkers, bigger dreamers, and bigger players will help move you towards success.

58. **Uncomfortable.** Successful people are willing to put themselves in new and unfamiliar situations. Being willing to be uncomfortable—to step out of your comfort zone—is a sure sign you're on your way to success.

   Following are seven helpful ways to move out of your comfort zone and toward success:
   - Get a goal to push you.
   - Get a challenge to push you.
   - Get a deadline to push you.
   - Push yourself with self-discipline.
   - Get others to push you.
   - Get competition to push you.
   - Get a tormentor to push you and a mentor to support you.

59. **Forward thinking.** Successful people tend to see things so much differently than most people. Where most people see problems, they see opportunities. The most successful entrepreneurs all started with a great idea, but most importantly they had the mindset and vision to see that idea through. Successful people are thinkers, innovators and doers that blaze a trail. Entire industries have started from ideas others once thought were crazy. Some of our country's most successful entrepreneurs are the most ambitious and creative people.

60. **Have a "can do" attitude.** Successful people practice a "can do" approach in almost every situation with the outlook that no matter what, it can be done. They consistently use phrases like "We can do it," "Let's make it happen," "Let's work it out"—and they always maintain that a solution exists.

Successful people talk in terms of explanations, resolving issues and communicating challenges with a positive outlook.

61. **Have an "I will figure it out" mentality.** Successful people believe there is always a solution, and you can do the same. Even if you're not sure how to do something, the best answer is "I will figure it out"—not "I don't know." You can admit you're unfamiliar with something—as long as you immediately follow that admission with the promise that you will figure it out or find someone who will.

62. **Focus on opportunity.** Successful people see all situations—even problems and complaints—as opportunities. Where others see difficulty, successful individuals know that problems solved equal new products, services, customers—and probably financial success. Remember, success is overcoming a challenge. Therefore, you can't succeed without some kind of difficulty.

63. **Embrace challenges.** Whereas many people loathe challenges—and use them as reasons to sink further into indifference—successful individuals are compelled and invigorated by challenges. Challenges are the experiences that sharpen successful people's abilities. To achieve your goals, you have to get to a place where every challenge becomes motivation for you.

64. **Seek to solve problems.** Successful people have a problem-solving mentality and typically are rewarded for their effort. Keep in mind that the bigger the problems—and the more people who benefit from the solution—the more powerful your success will be. One of the fastest and best ways to separate yourself from the masses is to establish yourself as someone who makes situations better, not worse. Furthermore, most successful businesses and products are solutions to problems.

65. **Able to give and receive constructive feedback.** Successful people have the courage to accept and provide specific feedback and constructive coaching without getting defensive.

66. **Accountable.** When it comes to accountability, successful business people know the buck stops with them. Being accountable starts when you assume responsibility for the outcome of your actions. Taking responsibility and being accountable for decisions you make—or don't make—is the key to success.

67. **Conscientious and reliable.** Successful people deliver on their promises in a timely manner. They do what they say they will do. If they are not able to follow through as promised, they apologize and do whatever they can to make it right.

68. **Altruistic and considerate.** Successful people truly care about the welfare of others and demonstrate this concern often.

69. **Trustworthy.** Building trust is paramount to the success of a business and in personal relationships. Successful people know the value of building trust. They understand that people don't invest in companies; they invest in people. Whether they are pitching an idea to potential investors, convincing a banker to issue a line of credit, or interacting with people outside of business, successful people exemplify trustworthiness.

70. **Provide value to others.** Successful people add value to other people in some way. By doing this they also create success for themselves. This doesn't happen by accident; it is intentional. Creating greater value for others is an extremely powerful success strategy for several reasons.

- **Increase your own value**
  When you create value for others, you become more valuable to them. Think about this in regard to your own life. Every single person who is important to you is adding value to your life on some level.

  What kind of contribution? Personal relationships are typically prioritized according to their emotional value. Having that person in your life adds value to your life because of the personal and emotional connection. In business relationships the value is generally measured by its effect on your bottom line or on your ability to do business.

- **A Changed perspective**
  When we focus on creating more value for other people, our perspective completely shifts. We naturally start moving away from self-serving motives as we increasingly consider the welfare of others. This causes our own happiness and success—a byproduct of our efforts to contribute to the happiness and success of others.

  When we focus on adding value to the lives of others, it expands our world. When our focus is based on selfishness, our world shrinks. As our world expands, we become exposed to greater opportunities. Acting on those opportunities brings greater value into our lives and that contributes to our success. It's a total win-win situation.

- **Success in business**

  If you are an employee, find ways to become more valuable to your employer. Creating greater value at work adds to your employer's bottom line, and his appreciation may add to yours. If you are an entrepreneur, look for new ways to create value for those in your tribe and those with whom you do business.

  The point is this: focus on adding value first—not how it is going to come back to you. Don't think of adding value to the lives of others as bartering. It's not about making a trade. It is more like making a deposit into the community human growth account. We all make deposits, and we all extract benefits. We just never know exactly how or when that is going to happen.

One thing we do know, if you make enough deposits you can eventually live on the interest. So take advantage of every opportunity to make deposits in that community human growth account. By doing so, you will be creating success by creating value. It's one of those situations where everyone benefits.

Success is no accident. It requires hard work, perseverance, studying. learning, sacrifice, and cultivating the ability to communicate well with others. Most of of all, it can usually only be achieved at a high level through passion and/or through a love of doing it.

CHAPTER 4

Prioritize: The Art of Prioritizing and Getting More Accomplished

Importance of Prioritizing

Prioritizing is important in all aspects of our lives. Everyone has felt overwhelmed and frustrated with having too much to do in a short amount of time. This leads to problems at work and in our personal and social lives. Prioritizing helps us identify what is most important at any given moment and focus our attention, time, and energy on that. When we are aware of what we need to accomplish, we are better able to spend our time on the right things.

We have all experienced that lost feeling of having so much to do we don't know where to start. It has been stated that every minute of planning saves 10 minutes of work, so while we cannot add actual time to our day, we can increase the time available to us by prioritizing. Everyone has many things to do and often not enough time or resources to do them. Prioritizing is about making choices regarding what to do next based on what is important.

Benefits of Prioritizing

Prioritizing leads to achieving long-term goals. A few immediate benefits of prioritizing include less stress, getting more accomplished without having to redo tasks, more free time and less wasted time. Prioritizing allows us to order our tasks by what is most important and/or urgent. It is about making choices and effectively using our time and improving our relationships.

Prioritizing also helps us be more proactive than reactive by allowing us to deal with unexpected tasks successfully in two ways:

1. a new, unexpected task does not ruin plans if normal activities are on schedule;
2. if a new, unexpected task is going to keep us from completing another task, we already have tasks prioritized, making it easier to determine which task gets bumped—the last one on the priority list

Tips for Prioritizing

The two most important things to keep in mind when prioritizing are lists and decision-making. The more time you spend creating your lists, the more effective and efficient you will be. If your day-to-day life is out of control, it's almost impossible to think strategically or plan effectively. When you're feeling overwhelmed about how much you have to do, it is difficult to focus on ensuring your life and work are efficiently moving in the direction you want to go. Thus, the more reason to get control of your daily activities. The better control you have on your daily activities the more you will be able to plan and establish a strategic course of action to accomplish your goals. Maintaining daily control will free up mental energy and time blocks to think about the big stuff.

Define what being "done" looks like. Most of the tasks people keep on their to-do lists are "amorphous blobs of undoability"—commitments without any clear vision of what being "done" looks like. Everything you are working on should have a very clear stopping point—a point where you know you are done. If you don't know what that point looks like, you will find it very difficult to make progress, and you will continually feel overloaded. It is important to clarify what being done looks like (i.e. envision the task or project being finished). Always start with the end in mind.

There are four lists you need to utilize to prioritize your work and life:

1. You should start with an overarching list with everything you know you need to get done. Put everything on this list.
2. Next, create a list for the month. This should include tasks that need to be completed during the month. If tasks can be placed on a calendar by due dates, you are already prioritizing.
3. Move on to the weekly list. Look at your monthly goals and work backwards. The tasks that are most urgent and those on which other tasks are dependent, along with those that have a due date in the current week should be on your weekly list. This list will be a living document in that it will change as you move through your week and new items appear. It's a good idea to put this list on a white board, possibly weekly calendar, on your office wall or somewhere you will see it regularly throughout the day. This serves as a visual reminder of where you are on your list.
4. And, finally complete the daily list—tasks to do today. This list will also be a living document, changing with new information received. End each day reviewing this list and creating the daily list for the next day—some items will carryover if not completed today. Begin each day previewing your daily list, so you know where you are headed and how you are going to get there. Starting your day with a plan gives you confidence.

Prioritize the items on your weekly and daily lists as you enter them. Use your organizational skills to put the most important/urgent tasks at the top of your lists and the least important/urgent ones at the bottom. As stated above, many tasks will prioritize themselves based on due dates. As you work on these lists, you may also realize some tasks are not feasible or as important as first thought and can be removed. Do not get caught up in re-prioritizing your lists constantly; you will mark off and add to items on your list as needed, but prioritize them where they should be and leave them. Electronic calendars are great tools for this, as they allow you to build on lists and carry over items from one day to the next without having to rewrite them.

What is the next most important task? Below are considerations for prioritizing items that may not have prioritized themselves through due dates:

- **Beneficial**—assess the value of the task. The completion of some tasks has more benefit than the completion of others. Think return on investment here. The tasks with the best returns should be on the top of the list. The return may be financial, a step toward a bigger accomplishment, or personal. The goal is to get the highest return for your mental and physical investments.

- **Realistic**—be careful not to set impossible goals. Don't set yourself up for failure and disappointment as this defeats the purpose of prioritizing. In other words don't put 20 tasks on your weekly list that you know will take you two hours each to complete.

- **Flexible**—be open to change. This was briefly visited on a previous page. Things are going to come up unexpectedly; take them as they come and determine where they fall in your priorities. Having prioritized your lists at creation makes being flexible much easier. When something unexpected needs to be dealt with (new task, new project, budget cuts, etc.) you will just need to determine where it falls on your list as far as importance. If this new, unexpected task, project, or information means you won't be able to complete everything on your list, then the bottom task falls off. It was the least important/urgent anyway, so this is not catastrophic, and you won't waste time at this point determining which task gets bumped.

- **Focused**—Your focus should be on completing your prioritized tasks in order, efficiently and effectively. Don't get caught up in perfecting or spending too much time on one task, as this may throw your schedule off and hurt other tasks/projects. Be sure you do a good job on the task at hand and move on.

Setting priorities is a continuous decision-making process. It is important to plan monthly. If you only plan daily, you will perpetually be in crisis mode and will simply be prioritizing your problems, or which fires to put out first. Ensure you are objective when deciding which tasks are top priorities. The fun, easy projects will not always be top priority and always placing the tasks you enjoy least at the bottom of your priorities only increases your dislike for them, because you will be rushed and pressured to complete them.

When prioritizing consider what is important, what is urgent, what is both, and what is neither. What is both important and urgent should be at the top of the list. Important but not urgent tasks should have a date set for completion and placed on your lists accordingly. If it is possible, delegate urgent but not important tasks to someone else. If this is not feasible in your situation, set aside a block of time to handle these tasks, i.e., answering phones, reading mail, paying bills, sending invoices, etc. Tasks that are neither important nor urgent should not be dealt with until the prioritized tasks have been. With that being said, you should set aside personal time each day, weekly at a minimum. This could be time for exercise, relaxing, coffee with a friend—whatever you will benefit most from—and it should be prioritized on your list for your sanity and health. You will be more productive and happier if you make yourself stick to this.

When you have large projects, you should break them into smaller components and set milestone deadlines for each. Work backwards from the final project due date, determine how much time you estimate each step will take and put deadlines on your calendar. These milestones should be prioritized accordingly on your lists. This will keep you ahead of schedule, so you are not rushing at the end, and if anything new comes up, you can handle the new task and meet your ultimate due date on the project. Everything you are working on is effected if you get behind schedule on one project.

The Multitasking Myth

Multitasking consists of paying attention to more than one task at a time. It may seem as though you are getting two things done at once, but all you are really doing is switching the focus of your attention back and forth between different tasks. Our brains can only focus on one thing at a time.

Think of attention as the beam of a flashlight. No matter how hard we may try, we can never shine the flashlight on two separate objects simultaneously. All we can do is rapidly switch the beam back and forth between them.

Every time we switch between tasks we lose productivity, since our brain has to take a moment to re-load information regarding that on which we are trying to focus. Multitasking may make you feel more productive, but it is a misleading feeling.

Mental Work: Collect, Process, Organize, Do, and Review

Not all work is the same. Following are five separate phases of effective work. By approaching work in a logical order you can be much more efficient and productive.

- *Collecting* is the act of gathering inputs: resources, knowledge, and tasks. You will have a much easier time making use of your available inputs, if they are all in one place before you begin.
- *Processing* is the act of examining your inputs: what you can do with the resources at your disposal. This is where you start separating things according to your next step: tasks, projects, future plans, and reference information.
- *Organizing* means inputting the results of the processing phase in a system you trust, so you don't have to remember it all. For instance, tasks go on your to-do list, projects go on a projects list, future plans go into a tracking system, and reference information goes into a file or database you can access easily.

- *Doing* means working through the tasks you can accomplish right now.
- *Reviewing* means examining the results of your work, revising your strategy, and improving your systems for better results.

Projects and Tasks

A critical mistake most people make when keeping track of things to do is failing to separate tasks and projects. Projects and tasks are two different things and should be tracked separately.

A task is a simple activity, one that normally shouldn't require more than 90 minutes to complete and is not related/or interconnected with any other task. Once you complete a task, it's done (finished); although you may do additional tasks involving the same activity.

A project, on the other hand, is a set of interconnected tasks that are reliant on each other and typically reliant on other people performing other relative task as well. A project includes multiple tasks. A project is also temporary.

Since projects and tasks are two different things, it is best to keep track of them separately. Tasks can be prioritized on a 'to-do' list. Projects are best addressed by breaking them down into manageable components.

Focus on the Next Action

Projects have many steps and can be overwhelming in their complexity. The key to handling these projects is to not focus on everything that has to be done. Instead, focus on the very next action you need to do to move the project forward. It may be looking up a piece of information, making a phone call, or accomplishing a small task. Whatever it is, it will move you closer to completing the project. Focus on what you can do right now.

Prioritizing by Categorization

For those that are visual learners, the Stephen Covey time management grid is an effective method of

	URGENT	NOT URGENT
IMPORTANT	*Quadrant I:* Urgent & Important	*Quadrant II:* Not Urgent & Important
NOT IMPORTANT	*Quadrant III:* Urgent & Not Important	*Quadrant IV:* Not Urgent & Not Important

FIGURE 1.   source: Stephen Covey, *7 Habits of Highly Effective People*

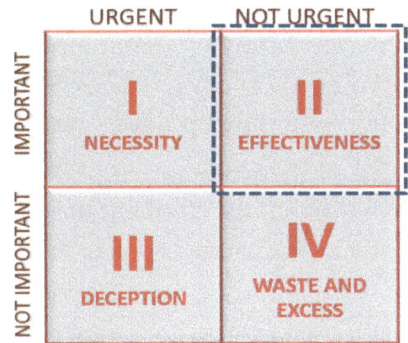

FIGURE 3.

	Urgent	Not Urgent
Important	**I** (MANAGE) • Crisis • Medical emergencies • Pressing problems • Deadline-driven projects • Last-minute preparations for scheduled activities Quadrant of Necessity	**II** (FOCUS) • Preparation/planning • Prevention • Values clarification • Exercise • Relationship-building • True recreation/relaxation Quadrant of Quality & Personal Leadership
Not Important	**III** (AVOID) • Interruptions, some calls • Some mail & reports • Some meetings • Many "pressing" matters • Many popular activities Quadrant of Deception	**IV** (AVOID) • Trivia, busywork • Junk mail • Some phone messages/email • Time wasters • Escape activities • Viewing mindless TV shows Quadrant of Waste

FIGURE 2.

organizing your priorities. As you can see from figure 1, there are four quadrants organized by urgency and importance.

• **Quadrant I** is for the immediate and important deadlines.

• **Quadrant II** is for long-term strategizing and development.

• **Quadrant III** is for time-pressured distractions. They are not really important, but someone wants them completed now.

• **Quadrant IV** is for those activities that yield little, if any, value. These are activities that are often completed when taking a break from time-pressured and important activities.

QUADRANT 1 Quadrant of Necessity ✓
Important, Urgent

Crisis
Emergencies
Disaster
Deadline Projects
Last Minute preparations
Pressing Problems

STRESSED LIFE! Stay away!!

QUADRANT 2 Quadrant of Quality ✓✓
Important, NOT Urgent

Exercise
Relationsip Building
Values Clarification
Book you want to write
Bonding with kids
Preparation.Planning

ACHIEVE GOALS FASTER!

QUADRANT 3 Quadrant of Deception ✗
Not Important, Urgent

Interruptions, some calls
mails, reports
Some meetings
"Pressing matters"
appear important but not important!

!!!!! Interrupts Productivity!

QUADRANT 4 Quadrant of Waste ✗
Not Important, Not Urgent

Facebook
Checking fun emails
Time wasters
Mindless TV Shows
Trivia, Busywork
Wasting good time

⚠ DANGER NO GOALS ACHIEVED!

FIGURE 4.

Many people find that most of their activities fall into quadrant I and III. Quadrant II is often under used; yet, it is exceptionally important, because one must work both tactically and strategically at the same time. Finding ways to expand Quadrant II activities is a common outcome from using this grid.

Figure 2 is another way of considering the grid.

To break it down even further, see figure 3.

And lastly, figure 4 is an example of how this priority grid may be viewed in more practical day-to-day terms.

The most obvious use of the grid is to take your current 'to-do' list and sort all the activities into the appropriate grid. Then, assess the amount of time you have to accomplish the lists and, if necessary, reallocate activities.

Bottom Line:
Do Important Things First!

Prioritizing for Balance

The above prioritizing tips are effective in both your work and personal life. Prioritizing encourages and induces better relationships. When you have prioritized your tasks, you will be more efficient at completing them and less stressed, which will allow more worry-free time to spend with family, friends, and hobbies.

It is important to mentally separate different areas of your life and devote your attention to each at the appropriate times. For instance, if you have prioritized your work tasks, you will be able to play with your children, spend time with your friends or spouse and devote your full attention to them. This will improve your relationships. When you are working you will be able to focus your full attention on the tasks at hand without worry regarding your personal relationships. This will improve your work.

Prioritizing Leads to Success

Prioritizing and sticking to it is the best ingredient for success. Stephen Covey, businessman and author of *The 7 Habits of Highly Effective People*, once stated, "Goals are pure fantasy unless you have a specific plan to achieve them." Warren Buffet, consistently recognized as one of the wealthiest and most generous of men, has been quoted as saying, "Don't risk what is important to you, to get what is not important to you." And, finally, Johann Wolfgang von Geothe summed up the essence of prioritizing when he stated, "Things which matter most must never be at the mercy of things which matter least."

Prioritizing helps you feel and be more in control of your life. Prioritizing also gives you a renewed sense of energy and excitement. Each day will go by faster and smoother helping you feel empowered and leading you to success—be it monetary, spiritual, or familial.

CHAPTER 5

Habits

A habit is a routine behavior repeated regularly and tends to occur subconsciously. It is a more or less fixed way of thinking, willing, or feeling acquired through previous repetition of a mental experience.

Some habits are known as "keystone habits," and these influence the formation of other habits. For example, the type of person who takes care of his/her body and is in the habit of exercising regularly will typically also be the type of person who eats better and uses credit cards less than people who do not. In business, safety can be a keystone habit that influences other habits which result in greater productivity.

The process by which new behaviors become automatic is called habit formation. Old habits are hard to break, and new habits are hard to form, because behavioral patterns we repeat are imprinted in our neural pathways; however, it is possible to break bad habits and form new habits.

Creating Positive Habits

Habit formation is the process by which a behavior, through regular repetition, becomes automatic or habitual. The process of forming a habit can be analyzed in three parts: the cue, the behavior, and the reward. The cue is the thing that causes the habit to come about, the trigger of the habitual behavior. This could be anything that one's mind associates with the habit and implores the mind to let the habit come to the surface. The behavior is the actual behavior one exhibits—the habit itself. Finally, the reward is the positive feeling after performing the behavior and continues the "habit loop." A habit may initially be triggered by a goal, but over time that goal becomes less necessary as the habit becomes more automatic.

Following are some proven steps to develop good habits.

1. Clearly identify the good habit you desire to form.
2. Make the decision to change and commit to the change.
3. Discover positive triggers for the habit.
4. Identify potential obstacles to the newly desired, positive habit.
5. Devise a plan that will help you develop and incorporate the new habit.
6. Employ visualization and affirmations. Visualization and affirmations are great for

integrating the new habit into your routine. While visualization is a powerful motivational tool and energizer, affirmations program the subconscious with the right mindset for establishing a new habit. Together they allow you to feel and imagine yourself carrying out the correct behaviors making it easier to adopt the new habit.

7. Enlist the support of family and friends along with any others who can help you develop the positive habit.

8. Find healthy ways to reward yourself.

9. Pray, Pray, Pray.

A variety of digital tools, including online and mobile apps, have been designed to support habit formation. Since there are so many types of habits you could build, it makes sense to find an app that best coincides with your specific habit-building goals. Following are several popular habit formation apps that may be of help to you:

- Evernote (Habit—Note taking)
- Coach.me (Habit—General)
- HabitRPG (Habit—General)
- HabitClock (Habit—Morning Routines)
- Loggr (Habit—Health & Fitness)
- The Fabulous—Habit & Routine (Habit—General Wellness)
- Chains.cc (Habit—General)
- Strides (Habit—General)

It is important to keep in mind that goals guide habits. A conscious goal pushes for an action, so the best approach to develop good habits is to establish clearly defined goals that align with your dreams and purpose. You should solidify your commitment to these goals by writing them out using the S.M.A.R.T. goals approach (addressed in more detail elsewhere in this book).

Breaking Bad Habits

Individuals possess a number of habits that can be classified as nervous habits, such as nail-biting. They are known as symptoms of an emotional state and are generally based upon conditions of anxiety, insecurity, inferiority and tension. These habits are often formed at a young age. When trying to overcome a nervous habit, it is important to resolve the cause of the nervous feeling rather than the symptom which is the habit itself.

A bad habit is an undesirable behavior pattern. Common examples include procrastination, fidgeting, overspending, stereotyping, and cracking knuckles. The sooner one recognizes these bad habits, the easier it is to correct them. A key factor in distinguishing a bad habit from an addiction or mental disease is willpower. If a person has control over the behavior, then it is a habit.

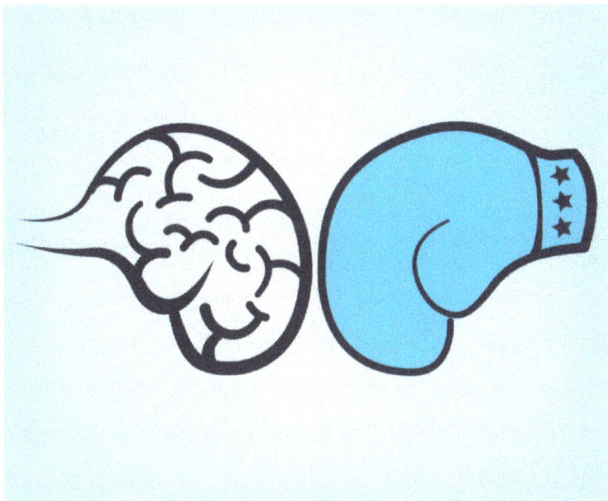

Many techniques exist for removing established bad habits, e.g., *withdrawal of reinforcers*—identifying and removing factors that trigger and reinforce the habit. Recognizing and eliminating bad habits as soon as possible is advised, as habit elimination typically becomes more difficult with age, because repetition reinforces habits cumulatively over one's lifespan. The key to changing habits is to identify your cue and modify your routine and reward.

Following are additional methods for breaking bad habits:

1. Take complete responsibility for your actions.
2. Scrutinize the consequences and rewards of your habits.
3. Weigh the short-term rewards with the long-term consequences.
4. Focus on breaking one habit at a time.
5. Don't take minor setbacks too seriously.
6. Track when you perform the bad habit.
7. Identify the triggers that lead to the bad habit (i.e. stress, boredom, etc.).
8. As much as possible, stop putting yourself in situations where the triggers to your bad habit flourish.
9. Replace your bad habit with a healthy habit.
10. Condition yourself to not enjoy the bad habit.
11. Find better alternatives that net the same reward as the bad habit.
12. Make a commitment to someone else. Tell others who will support you.
13. Surround yourself with people who live the way you want to live.
14. Break your timeline into manageable chunks—make it through the morning, day or week without engaging in your bad habit.
15. Visualize yourself succeeding.
16. Pray, Pray, Pray.

Length of Time to Develop a New Habit—Science vs. Myth

According to a survey conducted by psychologists at the University of Scranton, just eight percent of New Year's resolution-makers generally succeed in achieving their resolution (ouch). Judging from the most common resolutions in the survey, people really set themselves up to endure uphill battles. Losing weight, getting organized, spending less, and quitting smoking were at the top of the list. Any one of those could easily challenge one for an entire year.

According to U.S. News, approximately 80% of resolutions fail by the second week of February, which is approximately 45 days. If only these folks could have hung in there a little longer.

For the past several decades, it has been commonly believed that it takes 21 days to develop a new habit. The origins of this myth can be traced back to the 1950s. Plastic surgeon Maxwell Maltz noticed when he performed operations like nose jobs on his patients, it took those patients about 21 days to get used to seeing their new face. A similar phenomenon occurred when he performed amputations. Patients would sense a phantom limb for about 21 days after the operation. Lastly, he noticed that when he tried to form new habits for himself, it took him — you guessed it — about 21 days to do so. He then wrote about his experiences with this phenomenon in his 1960 book *Psych-Cybernetics,* with the precise thought being: "These, and many other commonly observed phenomena tend to show that it requires a minimum of about 21 days for an old mental image to dissolve and a new one to jell." Over time, that little tidbit became common knowledge — but with one crucial difference: The words "a minimum" got dropped. Hence: The myth of 21 days.

In a study released in the European Journal of Social Psychology, an Economic and Social Research Council postdoctoral team of researchers surveyed 96 people over a 12-week period to find exactly how long it takes to start a new habit. Each participant chose a dietary or activity behavior to be undertaken once daily. Then they proceeded to self-report how "automatic" the activity felt each time they did it. The researchers found that it took, on average, 66 days until the behavior reached peak levels of automaticity.

It's worth noting, though, that there was variation within each participant's results during the study; for example, it took one person only 18 days to reach peak automaticity, while another didn't get there at all by the end of the 84 days. Furthermore, some behaviors became habitual more easily than others. Simple behaviors reached automaticity quickly, while complex ones took longer. Their conclusion was that it takes an average of 66 days for a behavior to become a habit.

Habits of Unsuccessful People vs. Successful People

Achieving success in life is not something into which one falls ; success requires skills, timing, discipline, sacrifice, hard work and capitalizing on opportunities. Successful people share many common habits, which is enough to prove there is a recipe for success.

Some habits highly successful people share include tracking progress, learning from mistakes, persistence in goals, being humble, taking calculated risks, being organized, embracing changes and managing problems well. On the other hand, some habits common among unsuccessful people are wasting time, getting distracted easily, blaming others, not setting goals, being fearful of change, holding grudges, wanting others to fail, believing they know most everything, and negative thinking.

Following are the most common habits in which successful people engage.

1. Positive Thinking
Entertaining failure and doubting your beliefs is the first step of failing. However, positive thinking and surrounding yourself with positive-minded people are key ingredients to success.

2. Establish Clearly Defined Goals
Successful people have clearly defined short and long-term goals. Clearly defined goals are motivators within themselves.

3. Never Stop Learning
For successful people, every day is a learning day. Learning is something they enjoy and actively pursue.

4. Communicate Effectively
Effective communication is the glue that deepens your connections to others and improves teamwork, decision making, and problem solving. Effective communication allows you to convey negative or difficult messages without creating conflict or destroying trust.

Habits of 'Unsuccessful' People

Fear Change	Blame others
Act before they think	Think, say & do negative things
Try to bring others down	Get distracted easily
Give up easily	Do not have clearly defined goals
Think they know it all	Take the easy way out
Waste their time	Secretly hope another person(s) fail
Horde information & data	Hold a grudge
Talk more than they listen	Have a sense of entitlement
Aren't proactive in learning	Don't know what they want to be
Criticize, condemn, and complain	Frequently get angry at others

Habits of 'Successful' People

Have a burning desire	Strive to learn, improve & read everyday
Write down goals and targets	Accept responsibility for their failure
Work with passion & commitment	Spend time with the right people
Track progress	Maintain proper balance in life
Compliment others	Take calculated risks
Learn from their mistakes	Know purpose & mission
Think long-term	Forgive others
Are humble	Want others to succeed
Handle problems well	Talk about ideas
Make 'to-do' lists	Embrace change
Exude joy	Share information and ideas

5. Work Hard

Focused, hard work is the real key to success. The essence of being a hard worker is not only what you do when everybody is watching you but also what you do when you don't feel like it or when you are alone.

6. Take Risks

Some of the best moves successful people ever made involved taking risks. Taking risks is an integral part of business, although taking major risks without thinking them through is foolish. Risk implies the chance things may not work out the way we expect. However, there is a wide variety of benefits and advantages enjoyed by those who take calculated risks. Nothing ventured; nothing gained.

7. Proactive Disposition

Why bother setting goals, if you are not going to put forth the effort to achieve them? If you want to be successful, focus your energy on achieving your goals.

8. Intentional

Before successful people make any decisions, they weigh both sides of the coin. Being intentional means you are purposeful in word and action. It means you live a life that is meaningful and fulfilling to you. It means you make thoughtful choices in your life.

9. Not Controlled by Emotions

The worst thing you can do is to make important decisions based on emotions. Emotions fluctuate; therefore, emotionally charged times are not the best ones for decision making. Whether you are angry or happy, your decisions should be based on facts. Emotions influence people into making hasty decisions, which gets them into trouble or leaves them with feelings of regret. Not allowing your emotions to dictate your choices places you in a better position for success.

It is best to have a high emotional intelligence (EQ). EQ is the ability to identify and manage your own emotions and not allow the emotions of others to negatively impact you.

10. Integrity

Integrity means doing the right thing at all times and under all circumstances, whether or not anyone is watching. Honesty is always the best policy. If you want success, you must be honest in all your ways. It is also important to avoid those who are not trustworthy.

Habit Changes Successful People Make Before They Turn 35

Discovery and experimentation typically happen in the teen years. The 20s are a period in which people usually try to figure out who they are. The early 30s are often the years people begin to form a solid foundation for the rest of their life. Waiting until after 35 to make certain routine habit changes can result in some very detrimental consequences.

Recently, *Business Insider* compiled a list of things people should change in their lives to set themselves up for success. The data comes from business advisors, medical studies, articles, and user submissions. It's not a "who's who" list, but the advice is solid.

1. Stop Smoking

Research shows that those who quit smoking before 35 have a **90% lower mortality risk** than those who continue. A lower quality of life and an early death should be reason enough, right? Seek out support groups and professional assistance to kick the habit.

2. Stop Comparing Yourself to Others

Envy can be a motivator, but it is an unhealthy one. Step out of its ugly shadow and find your inner passion and motivation, rather than comparing yourself to others.

Envy's harmful effects:
- It fosters discontent and distress.
- It binds our freedom.
- It leads to resentment and bitterness.
- It causes us to do things we wouldn't normally do.
- It can spiral into depression.

Unfortunately, we live in an age where it's easier than ever to compare our achievements with those of others. Unfortunately, some social media outlets, such as LinkedIn and Facebook, advance the envy temptation bug all the more, causing sadness and depression.

Don't lie to yourself and think you're immune. Recognize and acknowledge it can happen to you and be prepared to fight it.

The biggest thing is to realize that Facebook, Twitter, Instagram and other social media shares are a "highlight reel" of that person's life. That person goes through life's struggles just as you do—maybe even worse. Learn to be happy for them and quit comparing your life to theirs. Everyone's circumstances, backgrounds, networks of influence, resources, and opportunities are different.

Instead of comparing yourself to others, focus on the things that motivate you. It's a better, healthier long-term lifestyle solution that provides a much better quality of life.

Consider these helpful, life-changing steps to overcoming envy:

1. Shift your focus to the goodness in your life.
2. Remind yourself that nobody has it all.
3. Avoid people who habitually value the wrong things.
4. Spend time with grateful people.
5. Understand that marketers routinely fan the flame.
6. Celebrate the success of others.

Happiness replaces envy, and there is plenty enough of that to go around.

3. Stop Trying to Satisfy Everyone

Saying "No" is often difficult. It requires you to cut connections, own up to not being able to satisfy everyone, potentially create an adversary, or even sever a relationship; nevertheless, it is an essential step towards success.

In addition to saying "No," reducing demands on your time can also be obtained by culling your social networks. Time management experts and professional life coaches recommend whittling down your social network list to only those people from whom you truly want to hear.

4. Sticking to a Healthy Sleep Routine

Sleep is critical to success. Sleep also plays an important role in your physical health. For example, sleep is involved in healing and repairing your heart and blood vessels. Ongoing sleep deficiency is linked to an increased risk of heart disease, kidney disease, high blood pressure, diabetes, and stroke. Going without sleep for a single night causes changes in the brain similar to a knock in the head. According to Harvard Medical School, one night without enough sleep can cause elevated blood pressure throughout the next day.

Following are some helpful ideas to facilitate better sleeping habits:

1. The apps SleepCycle for iOS and Sleep as Android for Android analyze your sleep. When it's time, they wake you up in your lightest sleep phase.
2. Try to avoid all electronic devices and the television prior to bedtime. The American Medical Association's Council on Science and Public Health research has concluded that exposure to excessive light at night, including extended use of various electronic media, can disrupt sleep or exacerbate sleep disorders.
3. Filter your phone's blue light for less distraction, so you can get a better night's sleep. Cell phones, computer screens, and tablets trigger your brain to stay awake for quite some time. Blue light at night — the kind given off by electronic devices — is a bad thing. It effectively tricks your brain into thinking it is still daytime. It suppresses the secretion of melatonin, a hormone produced at night to prepare the body for sleep.

4. Try to go to sleep and get up at the same time every day.

5. Get as much natural sunlight as possible during the day.

6. Move vigorously during the day—don't sit for more than an hour.

7. Limit caffeine, nicotine, alcohol, and big meals at night.

8. Take time for relaxing activities before sleep.

9. Create a calm and restful sleep environment.

10. Control your exposure to light. Make sure the room is dark.

11. Keep the lights down if you get up during the night.

12. Keep noise down.

13. Cool your room. Most people sleep best in a slightly cool room (around 65° F or 18° C) with adequate ventilation. A bedroom that is too hot or too cold can interfere with quality sleep.

14. Make sure your bed is comfortable.

Bedtime rituals to help you relax

Create a "toolbox" of relaxing bedtime rituals to help you unwind before sleep:

- Read a book or magazine in soft light
- Take a warm bath
- Listen to soft music
- Perform some easy stretches
- Wind down with a favorite hobby
- Listen to books on tape
- Make simple preparations for the next day
- Dim the lights in the hours leading up to bedtime

If all of the above does not help, or if you suffer from insomnia, contact the National Sleep Foundation www.sleepfoundation.org

5. Regular Exercise and Healthier Dietary Choices

If you take care of your body, it will take care of you, and you will find it much easier to pursue and fulfill your dreams. As we age, our body deteriorates. Regular exercise and healthy eating is vital for slowing the deterioration, staying fit, enjoying a better quality of life, and achieving your goals.

6. Save Money

The spontaneous and easy-spending ways need to stop, hopefully way before the age of 35. Saving money can help you become financially secure, provide you

funds for prudent investments that will earn you more money, and give you a safety net in case of an emergency.

Ideally, on top of saving money, it is best to get rid of debt. Debt typically only makes sense if it is for something that will increase in value and substantially offset all the negatives associated with debt.

If you're new to the idea of saving money, implement the following measures in your life:

1. Set goals
2. Budget
3. Negotiate prices
4. Slash excess spending
5. Cut monthly bills
6. Make smarter grocery choices

7. Transfer debt to a lower interest credit card

8. Find savings and checking accounts that provide the best value and lowest cost

9. Invest under the right tax shelter

10. Consider owning a different car

11. Shop sales. However only buy the things you absolutely need and can't get elsewhere at a better price (borrow, used).

12. Make alternative choices (eat at home rather than eating out)

13. Switch your wireless provider

14. Buy prescriptions (at a cheaper price) at a different location or online

15. Use an automated savings plan with direct deposits

16. Reduce your energy and utility costs

17. Re-shop your car insurance

18. Buy a coffee machine; or better yet, give up caffeine

19. Track your spending

20. Make purchases only when necessary. Don't buy items that can be borrowed. Never make purchases just because they are on sale. It is only a good deal if you need it.

7. Start Pursuing a Dream

Everyone has a dream or passion. Certainly, by age 35 you need to be pursuing your dream.

However, it is not just enough to have a defined dream. To reach your dream as quickly and efficiently as possible the dream must be clearly defined and transformed into a measurable goal—meaning obtaining the goal can be validated.

The goal then needs to be broken down into manageable increments through a S.M.A.R.T. goal process as addressed in more detail in other parts of this book. Your goals also need to be arranged in priority order. Nothing says you can't pursue more than one goal at a time, but if given limited resources you know what to spend your time and efforts on.

It is also very important that you step back occasionally and evaluate your progress. Are your plans and actions bringing you closer to reaching your goals? Are you any closer today than you were when you first started? Are your goals as measurable, realistic, and clear as you once thought? If not, then something isn't working, and that calls for change.

8. Stop Letting the Past Define Your Future

Success is about the process of learning from your mistakes and failures—not dwelling on them, beating yourself up, or blaming other people. By 35 you need to forgive yourself, forgive others, and put past mistakes behind you in order to pursue a good quality of life and reach your goals.

Prayer, self-improvement books, and professional help are great resources to assist in overcoming issues in the past that may be hindering you. Don't let the past be a stumbling block to your present level of happiness and achieving future success.

'Must Do'—Positive Habits to Develop

Quite often we get so caught up in day-to-day activities we get distracted from our dreams and purpose. Following are positive habits you should incorporate into your routine for a more purposeful and better quality life.

1. **Identify your dreams and purpose.** Write down clearly defined goals that align with your dreams and purpose in life. Pursue your goals via the S.M.A.R.T. goal method.

2. **Start the day with a positive mind-set.** Upon awakening make the commitment to maintain a positive attitude regardless of what happens throughout the day. God is in control, and everything happens for a reason. Thinking positively is the only way you will be able to function at an

optimal level and deal with things in the best way possible.

3. **Practice Gratitude.** Be thankful in all circumstances and for all things; this will bring you joy and peace. Keep in mind that often the hardships and most challenging times of life bring about the most positive changes and tremendous personal growth.

4. **Learn something new.** Make a conscious effort to keep your brain active and functioning at optimum levels. Make every effort to continue learning, as this will keep you sharp, make you a better person, and potentially lead to significant rewards.

5. **Have a good laugh.** Laughter and a jovial attitude help relieve stress, makes life much more enjoyable, and keeps us from taking ourselves too seriously.

6. **Smile at someone.** A winner acknowledges others and gives them a smile. Giving others a smile will make you and the other person feel better. A simple smile can also break down barriers to connecting with others and attract opportunities.

7. **Tell and remind others how much you appreciate them; and give them genuine compliments.** Everyone appreciates a nice compliment and enjoys positive feedback. Quite often we take others, and especially those to whom we are closest, for granted. Telling others we care about them or cherish our relationship with them is incredibly enriching on many levels. Giving others sincere compliments not only positively impacts those being complimented, but helps us feel better, too. Telling others we appreciate them and giving sincere, valid compliments are also key ingredients to developing and strengthening relationships.

8. **Perform an act of kindness.** Do something nice just for the sake of doing it. Help others regularly. This will generate and promote good will.

9. **Be a better listener.** Take the time to actively listen to others' points of view. Even if you don't agree with what is being said, or find it boring, try to put yourself in their place and understand their point of view

10. **Take quiet time(s).** Make a habit to regularly give yourself a break from all stimuli to process, relax, and pray. Even a little 15-minute catnap can be surprisingly refreshing and rejuvenating.

Habits—The Last Word

It is fairly easy to identify the bad habits and cast blame for our failures and shortcomings on them, but rarely do we attribute our successes to our good habits. Nevertheless, habits can MAKE you, but they can also BREAK you. Good or bad, they are what define you about as much as anything else in life, including your DNA. They make you who you are or they break you down, and keep you from becoming all that you could be. Therefore, in order to get the maximum benefits out of life breaking bad habits and developing good habits needs to be one of our highest priorities.

CHAPTER 6

Time Management

Time Management refers to managing time effectively so that the right time is allocated to the right activity. Effective time management allows individuals to assign specific time slots to activities per their importance. **Time Management means making the best use of time** as time is always limited.

What is the importance of time management in your life and work? How much does being able to manage your time well actually matter?

Before we can even begin to manage time, we must learn what time is. A dictionary defines time as "the point or period at which things occur." Put simply, time is when stuff happens.

There are two types of time: clock time and real time. In clock time, there are 60 seconds in a minute, 60 minutes in an hour, 24 hours in a day and 365 days in a year. All time passes equally. When someone turns 50, they are exactly 50 years old, no more or no less.

In real time, all time is relative. Time flies or drags depending on what you're doing. Two hours at the department of motor vehicles can feel like 12 years. And yet our 12-year-old children seem to have grown up in only two hours.

Which time describes the world in which you really live — real time or clock time?

Time management gadgets and systems are designed to manage clock time. However, we live in real time, a world in which all time flies when you are having fun or drags when you are doing your taxes.

The good news is that real time is mental. It exists between your ears. You create it. Anything you create, you can manage. It's time to remove any self-sabotage or self-limitations you have around "not having enough time," or today not being "the right time" to start a business or manage your current business properly.

There are only three ways to spend our waking time: thoughts, conversations and actions. Regardless of the business, your work will be composed of those three items.

Any of the three may be frequently interrupted or pulled in different directions. While you cannot eliminate interruptions, you can manage how much time you spend on them and how much time you spend on the thoughts, conversations and actions that will lead you to success.

Why Properly Managing Our Time Is Important

Whether we assign a dollar value to it or not, time is valuable to us. Think about it: How much of your typical work week do you spend stressed about not having enough time to complete a task or reach a goal?

There are only so many hours in a day, none of which can be reclaimed. How many hours a day do you have left today? Whatever your definition of time management, it can't be stored, saved or borrowed. Once it's gone, it's gone. Time management is about making the most of your time — and the more you value it the better you'll use it.

There are various ways to tackle the issue of time management — you can download apps, adjust your sleep time, create lists, etc. However, if you don't fully understand why it's important for you to better manage your time, those apps and lists aren't going to help you. If you don't have the motivation to use them, you won't.

To appreciate the importance of managing time effectively and efficiently you first must get a good understanding as to what you stand to gain from it. To obtain the motivation and discipline needed to develop better time management skills, review the following 20 reasons why time management is so important.

1. **Time is limited** — No matter how you slice it, there are only 24 hours in a day. If you want to accomplish more and achieve greater success you have to recognize that time is limited and acknowledge the importance of effectively managing this limited resource.

2. **You can accomplish more with less effort** — When you learn to take control of your time, you improve your ability to focus. And with increased focus comes enhanced efficiency, because you don't lose momentum. You will breeze through tasks more quickly.

3. **Improved decision-making ability** — When you feel pressed for time and have to make a decision, you're more likely to jump to conclusions without fully considering every option. That leads to poor decision making.

   Through effective time management, you can eliminate the pressure that comes from feeling like you don't have enough time. You'll start to feel calmer, more confident, and in control. When it is necessary to examine options and make a decision, you are able to do so with clearer thinking, which diminishes the chances of making a bad decision.

4. **Become more successful in your career** — Time management is the key to success. It allows you to take control of your life rather than following the flow of others. As you accomplish more each day, make more sound decisions, and feel more in control, people notice. You gain more respect and admiration from others as someone on whom they can count to get things done, which increases success and the opportunity for advancement.

5. **You learn more** — The more you learn, the more likely you are to succeed. Great learning opportunities are all around you (others, Internet, library,

education system, organizations/clubs, business owners, etc.).

When you work more efficiently, you have that time to better educate yourself and pursue professional development opportunities. You can volunteer or even pick up a part-time job where you can hone your skills and/or identify your passion. There are more opportunities to have lunch with people who can assist or mentor you.

The more you learn about your company and your industry, the better your opportunities for success. Effective time management will provide you the time to learn and grow.

6. **Reduce stress** — When you don't have control of your time, it's easy to end up feeling rushed and overwhelmed. When that happens, it can be hard to figure out how long it will take to complete a task. Learning to manage time efficiently will reduce the amount of unhealthy stress you feel.

Once you learn how to manage your time, you no longer subject yourself to that level of stress. Aside from it being better for your health, it will also give you a clearer picture of the demands on your time, which will allow you to better estimate how long a given task will take you to complete, and know whether or not you can meet a deadline.

7. **Free time is necessary** — Everyone needs time to relax and unwind. Unfortunately, though, many of us don't get enough of it. Between jobs, family responsibilities, errands, and upkeep on the house and yard, most of us are hard-pressed to find even 10 minutes to sit and do nothing.

Having good time management skills helps you find that time. When you're busy, you're getting more done. You accumulate extra time throughout your day that can later be used to relax, unwind, and prepare for a good night's sleep.

8. **Self-discipline is valuable** — When you practice good time management, you leave no room for procrastination. The better you get at it, the more self-discipline you learn. This is a valuable skill that will begin to impact other areas of your life where a lack of discipline may have kept you from achieving a goal.

9. **You're more fulfilled** — People often think getting organized means time management software, lists, planners and diaries, but it goes beyond that. It starts with the choices and decisions you make based on the values you hold. When you know what matters and handle those tasks efficiently, it's time well spent. How you function affects how you feel about the whole of your life.

10. **You have more energy** — Strange but true — the act of finishing tasks often brings a level of satisfaction and energy that makes you feel good. The importance of time management here? It will help you do more of those endorphin-releasing activities. Your ability to manage time has a direct affect on your energy levels.

11. **You develop more qualities** — Once you apply skills, techniques and strategies, you'll find they work in conjunction with qualities we all have, but don't all use: Patience, persistence, self-discipline and assertiveness will all develop. As you expand your awareness of time, your ability to manage it improves, too.

12. **You achieve what you want to and need to faster** — Better time management means you accomplish more.

13. **You enjoy your life more** — The more value you put on your time, the greater your ability to learn how to do what matters, allowing you to enjoy life more.

14. **More Free Time** — While you can't create more time, but you can make better use of it by efficiently managing it. Even simple actions like shifting your commute or getting your work done early can produce more leisure time in your life.

15. **Less Wasted Time** — When you know what you need to do, you waste less time in idle activities. Instead of wondering what you should be doing next, you can already be a step ahead in your work.

16. **More Opportunities** — Being on top of your time and work produces more opportunities. The early bird always has more options, and luck favors the prepared.

17. **Improves Your Reputation** — Your time management reputation will precede you. At work and in life you will be known as reliable. No one is going to question whether you are going to show up, do what you say you are going to do, or meet that deadline.

18. **Less Effort** — A common misconception is that time management takes *extra* effort. To the contrary, proper time management makes your life easier. For example, things such as packing for a trip or completing a project take less effort.

19. **Get More Done** — Of course, being productive is one of the main goals of time management. When you are aware of what you need to do, you are better able to manage your workload. You will be able to get more (of the right tasks) done in less time.

20. **Less Rework** — Being organized results in less rework and mistakes. Forgotten items, details, and instructions lead to extra work. How often do you have to do a task more than once? Or make an extra trip because you forget something? Being organized lessens the frequency of these.

The Positive Cycle of Good Time Management

Looking through the list above, it's easy to see the multiplicative effect of time management. Good time management allows you to accomplish more in a shorter period of time, which leads to more free time, which lets you take advantage of learning opportunities, lowers your stress, and helps you focus, which leads to greater professional success. Each benefit of time management improves another aspect of your life. Effective overall time management benefits you in *all* areas of your life.

The importance of time management depends on the value you place on your time. It can also be argued that the importance of managing time effectively and efficiently is dependent upon the value you place on your life and what you desire to accomplish with your life. Developing self awareness of the value of time in your life is the first step. Learning time management skills is the next.

Time Management Improvements

"Time management" is the process of organizing and planning how to divide your time between specific activities. Good time management enables you to work smarter — not harder — so you get more done in less time, even when time is tight and pressure is high.

The highest achievers manage their time exceptionally well. By using the time-management techniques in this section, you can improve your ability to function more effectively under even the most adverse conditions.

Good time management requires an important shift in focus from activities to results. It is important to realize that b**eing busy isn't the same as being effective.**

Spending your day in a frenzy of activity often achieves less, because you're dividing your attention between so many different tasks. Good time management allows you to focus on the project, so you get more done in less time.

Ways to Improve Your Time Management Skills

Do you often feel stressed out due to too much work? Do you feel you have more tasks on hand than time

to do them or do you feel with effective use of your time you could complete all the given tasks?

The trick is to organize your tasks and use your time effectively to get more things done each day. This can help you reduce stress and increase production. Time management is a skill that takes time to develop and is different for each person. You have to find what works best for you. Below are 17 proven strategies that may help you.

1. **Delegate Tasks:** It is common for all of us to take on more tasks than we can comfortably handle. This can often result in stress and burnout. Delegation is not running away from your responsibilities but an important function of management. Learn the art of delegating work to your subordinates as per their skills and abilities.

2. **Prioritize Work:** Before the start of the day, make a list of tasks that need your immediate attention — too often unimportant tasks consume an inordinate and unnecessary amount of your precious time. Some tasks need to be completed immediately on a specified day, while other less pressing tasks could be carried forward to another

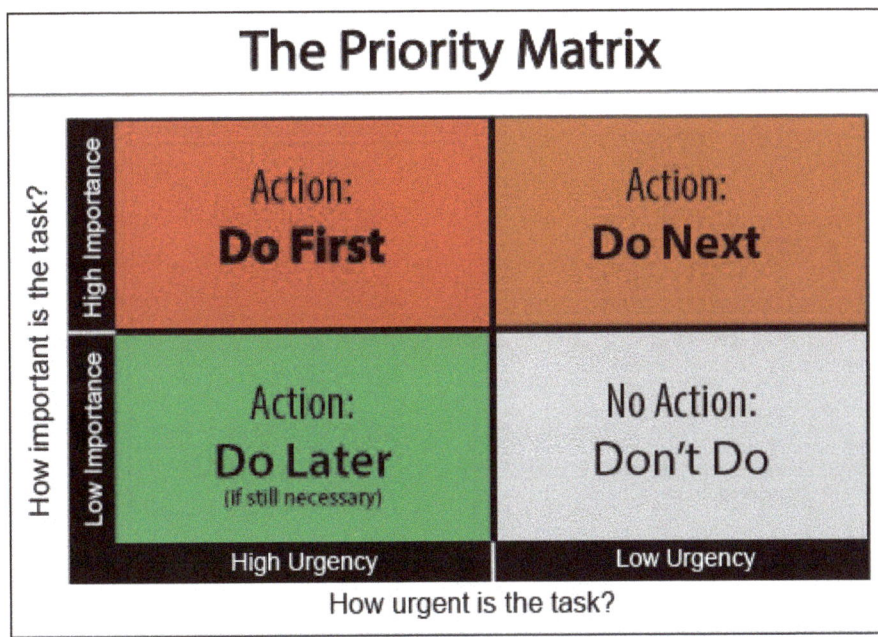

The Priority Matrix

	High Urgency	Low Urgency
High Importance	Action: **Do First**	Action: **Do Next**
Low Importance	Action: **Do Later** (if still necessary)	No Action: Don't Do

How important is the task? / How urgent is the task?

day. In summary, prioritize your tasks to focus on those that are most important.

3. **Avoid Procrastination:** Procrastination negatively affects productivity. It results in wasting essential time and energy and should be avoided at all costs. It will become a major problem in one's career and personal life if not addressed. Even small amounts of time lost here or there, due to procrastination, can add up to a significant loss of time in a short period.

4. **Schedule Tasks:** Carry a planner or notebook with you and list all tasks that come to your mind. Make a 'To Do' list before the start of each day, prioritize the tasks, and make sure each is attainable. High achievers typically make their 'to do' list out for the next day at the end of each day. To better manage your time management skills, you may find it beneficial to make three 'to do' lists: work, home/family, and personal.

5. **Avoid Stress:** Stress often occurs when we accept more work than is within our ability to reasonably achieve. The result is that our body starts feeling tired which can affect our productivity. Instead, delegate tasks to your juniors and make sure to allow time for relaxation.

6. **Set Deadlines:** When you have a task at hand, set a realistic deadline and stick to it. Try to set the deadline a few days before the task is actually due, so you have time to deal with all those tasks that may interrupt you along the way. Challenge yourself and meet the deadline. Reward yourself for meeting difficult challenges.

7. **Avoid Multitasking:** Most of us feel that multitasking is an efficient way of getting things done, but the truth is we do better when we focus and concentrate on one thing. Multitasking hampers productivity and should be avoided to improve time management skills.

You are not doing yourself, your company, or your friends and family any favors by multitasking. Research shows it's not nearly as efficient as we like to believe and can even be harmful to our health.

When it comes to attention and productivity, our brains can only handle a finite amount. It's like a pie chart, and whatever we're working on is going to take up the majority of that pie. There's not a lot left over for other things, with the exception of automatic behaviors like walking or chewing gum.

Moving back and forth between several tasks actually wastes productivity, because our attention is expended on the act of switching gears. Furthermore, it is almost impossible to get fully "in the zone" of any one activity when we're multitasking.

Contrary to popular belief, multitasking doesn't save time. In fact, research shows that it typically takes longer to finish two projects when jumping back and forth than it does to finish each one separately. According to a study published by the American Psychological Association, switching between tasks can cause a 40% loss in productivity.

8. **Start Early:** Most successful men and women have one thing in common: they begin their tasks for the day early. As the day progresses, energy levels start going down which negatively affects productivity and performance.

9. **Take Some Breaks:** Take a break occasionally. Too much stress can take a toll on your body and affect your productivity. Take a walk, listen to music or do some quick stretches. Or, you could take time off from work to spend time with your friends and family and/or do something you love. Breaks refresh and reenergize us.

10. **Learn to say No**: Politely refuse to accept additional tasks, if you think you are already overloaded with work. Take a look at your 'To Do' list before agreeing to take on extra work.

11. **Allocate time allowed.** Activities, and even conversations, should have an assigned time period. Appointment books work. Schedule appointments with time frames for yourself and your 'to-do' list items. Create time blocks for high-priority thoughts, conversations, and actions. Schedule when activities will begin and end and have the discipline to keep these appointments.

12. **Live the 80/20 Rule.** The 80/20 Rule is one of the most helpful of all concepts of time and life management. It is also called the Pareto Principle after its founder, the Italian economist Vilfredo Pareto, who first wrote about it in 1895.

This rule says that 20% of your activities will account for 80% of your results. 20% of your customers will account for 80% of your sales. 20% of your products or services will account for 80% of your profits. 20% of your tasks will account for 80% of the value of what you do and so on. This means that if you have a list of ten items to do, two of those items will turn out to be worth as much or more than the other eight items put together.

Make a list of all the key goals, activities, projects and responsibilities in your life today. Which of them are, or could be, in the top 10% or 20% of tasks that represent, or could represent, 80% or 90% of your results?

Resolve today that you are going to spend more and more of your time working in those few areas that can really make a difference in your life and career and less and less time on lower value activities.

13. **Schedule time for interruptions.** Plan time to be pulled away from what you're doing. Build extra time into each of your appointments (meetings, thoughts, tasks as listed above) to allow for the inevitable interruptions that will come your way.

14. **Think about it.** Immediately before every call and task decide what result you want to attain. This will help you know what success looks like before you start. Following each call and activity consider whether or not your desired result was achieved. If not, what was missing? How do you put what's missing in your next call or activity?

15. **Do not disturb.** Put up a "DO NOT DISTURB" sign when you absolutely have to get work done.

16. **Take control of your response to incoming.** Just because the telephone rings doesn't mean you have to answer it. Same goes with text messages. Disconnect all instant messaging, including social media, news feeds, and financial notifications. Don't instantly give incoming your attention, unless it's absolutely crucial in your business to offer an immediate human response. Instead, schedule a time to answer or return phone calls, return messages, and reply to emails.

17. **Block out Social Media distractions.** Distractions like Facebook and other forms of social media, unless you use these mediums as tools to generate business, are a major time drain and should be avoided. Even when using these mediums for your business, be mindful you don't allow them to distract you from the task at hand.

Time Management Tips for Dealing with Email

1. **Add "No Reply Needed."** If appropriate, insert "No Reply Needed" in the "Subject" line or the opening of your email. This can reduce the number of return emails you receive.

2. **Save time by using pre-written responses.** Use pre-written responses to frequently asked questions or requests for information, such as directions, fee schedules, or "how-to" guidelines. Then you can cut-and-paste your reply emails, saving time.

3. **Save time by answering under the question.** When someone send an email asking several questions, reply with, "See my response in bold text under your questions below." Then simply insert your responses after their original questions in bold text.

4. **Attach first, write next.** If you're sending an attached file, attach it when you first start writing the email, so you don't forget to do it.

5. **Use a clear subject line.** Put enough details in the subject line so recipients know the gist of your email right away. This will help reduce replies and questions, while also reminding you immediately of the subject, when there is a reply.

6. **Keep email focused on one topic.** Have one email per subject. People often respond to your first and last questions, but overlook or forget the others, so keep things simple. Use one email to address the meeting reminder, for example, another the department social event, and another for the status of a particular report. Recipients can respond accordingly, as they have the time and necessary information. This also saves you time in organizing your email, prioritizing your responses, and reading responses.

7. **Use the telephone when appropriate.** A phone call is sometimes faster and easier than sending numerous back and forth emails; a phone call is also considered a more personal touch in some situations.

Checking Email

Checking your email regularly during the day can be an effective way to keep your inbox at manageable levels. However, the constant interruption and distraction that comes from multitasking in this way can dramatically lower your productivity and disrupt your ability to enter a state of flow when working on high value projects.

One strategy you can use is to check email only at set points during the day. For instance, you may decide you'll only check your email first thing in the morning, before lunch, and at the end of the day. It helps to set your email software to "receive" messages only at certain times, so you aren't distracted by incoming messages. If you can't do this, at least make sure you turn off audible and visual alerts.

You can also reserve time to read and respond to email after a long period of focused work, or at the time of day when your energy and creativity are at their lowest (this means you can do higher value work at other times).

If you're concerned that your colleagues, boss, or clients will be annoyed or confused that you're not responding to their emails quickly, honestly explain that you only check email at certain times as a means of better managing your time and being more productive. Be sure to let them know they are welcome to call you for urgent matters.

Time Management Tips When You're Drowning in Email

Keep your email box as lean as possible. Having too many emails in your in-box is like having a pile of papers on your desk. It can be stressful just thinking about all that "stuff" with which you have to deal. Instead, organize your email with folders for each project, client, and/or subject. Taking to time organize your email this way will save time in the future. Don't use your 'In-box' or 'Sent' folder as a huge miscellaneous file. If you keep things organized in your email program, you'll also keep them better organized in your mind

Following are a few more ideas to better manage email to save time:

1. Decide what action or response is necessary immediately after reading an email. By deciding right away, you save time not having to reread and rethink it. Make it a priority to handle each email only once when possible.

2. Once you've read an email, make one of these four decisions.

    Option #1: Dump it. Delete the email, unless you have a good reason not to do so.

    Option #2. Delegate it. Forward the email to someone else and have them deal with it.

    Option #3. Delay it. Postpone your response, if you're waiting for more information, if it is a time-consuming issue that is not urgent, or if it is a trivial request. If the sender resends the request, apologize and do what is prudent to promptly end the matter.

    Option #4. Do it. Immediately do what the email requests or requires.

3. Use the "Tools/Organize" or "Tools/Rules" function in your e-mail program to color-code

incoming email from key people, so those emails stand out from others.

4. Create an "Action Items" folder for important email that needs attention. As stated above, you want to deal with emails immediately after reading; however, some may require more information or action that must be dealt with at a later time. Review the items in this file daily and save them to your hard drive or delete them when you've finished them.

5. Glance at all new email "Subject" lines and delete the junk mail as you go. As you do this, look for the important ones that you'll read.

6. Use the "Block Sender" option to prevent emails from specific individuals or companies that are always nothing more than a waste of time from cluttering your inbox.

7. Unsubscribe from solicitation emails.

8. If your email system can organize messages according to "threads," read the last message first in a thread that deals with a particular subject. Many times you won't need to read the previous ones.

9. Before you set up auto-filing features consider whether urgent mail might wind up being auto-filed before you see it.

10. Avoid getting on lists for jokes, cute stories, etc. If you like to receive this kind of material, set up an auto-filing function to send them into special files you can review at your leisure.

11. Check your email only at designated times. You should attempt to work until you come to some kind of natural break, a stopping point, or a designated time.

12. If you have to keep complete records of email correspondence, save your "Reply" e-mail. When you reply to people's email, a copy of their entire e-mail is automatically included in the reply.

13. Use one address if you register for something on the Internet (which might attract spam), another for business, and another for personal use.

14. Regularly purge your email of outdated and unnecessary messages. Archive e-mail you need to keep for historical reasons.

15. If you don't want everyone in a group to see each other's e-mail addresses and desire to prevent having to sort through a bunch of 'back-and-forth' traffic, send the e-mail as a blind copy (BCC). BCC will keep the email private.

16. When replying to sender only, don't do a "reply all." Don't "reply all" to an entire group, if your message is not relevant to everyone.

17. If you forward a message, put your comments at the top rather than at the end. This will save confusion and reduce reply questions.

Understanding the Difference between "Urgent" and "Important"

- *"Urgent"* tasks demand your immediate attention.
- *"Important"* tasks matter and not doing them may have serious consequences for you or others. For example:

- **Getting gas when near empty is urgent**. If you don't do it, there will be significant, immediate consequences.
- **Brushing your teeth regularly is important** If you don't, you may get gum disease, cavities, or develop other problems, but it's not urgent. If you leave it too long; however, it may become urgent.
- **Picking your children up from school is both urgent and important.** If you are not there at the right time, they will be waiting in the playground or the classroom.
- **Reading funny emails or checking Facebook is neither urgent nor important.** A significant amount of valuable time can be wasted on these activities. Even doing so for only a few minutes during the day can add up to a lot of wasted time over a week, month and year. These time drains distract you from getting your urgent and important tasks done and get in the way of your goals.

This distinction between urgent and important is the key to prioritizing your time and your workload, whether at work or at home. To better get a handle on where items best "fit" in your life try using a grid, like the priority matrix below, to organize your tasks into their appropriate categories:

Remember, also, that your health is important. Just because you have lots of very important things to do doesn't mean you should avoid exercising, preparing a healthy meal, or even taking time for 10-minute walk. You should not ignore your physical or mental health in favor of more "urgent" activities.

Furthermore, urgency and/or importance is not a fixed status. You should review your task list regularly to make sure nothing should be moved up or downgraded, because it has become more or less urgent and/or important.

What can you do if an important task continually gets bumped down the list by more urgent but still important tasks?

First, consider whether it is genuinely important. Does it actually need doing at all, or have you just been telling yourself that you should do it?

If it really is important, then consider delegating it, if possible, or try to make other arrangements (i.e. use automatic bill pay instead of mailing a check each month).

Further Principles of Good Time Management

Clutter can be both a real distraction and genuinely depressing. Tidying up can improve one's sense of both self-worth and motivation. It is also easier to stay on top of things if your workspace is tidy.

Top Tip for Tidying:

Create three piles: Keep, Give Away, and Throw Away.

1. **Keep**, if you need to keep it for your records, or do something with it. If it needs action, add it to your task list.
2. **Give away**, if you don't need it, but someone else might be able to use it. This also includes issues that can and should be delegated.
3. **Throw away** (or recycle) things that have no value to you or anyone else.

Know your Flex, Peak and Weak Times of the Day

Most of us have times of day where we are more alert and productive than other parts of the day. It is best to schedule the highest priority items for the times in which we have the most energy. Other items like meetings, errands, and basic administrative responsibilities can be scheduled during other times of the day.

Another useful option is to have a list of important but non-urgent, small tasks that can be done in that odd 10 minutes we have available here and there throughout the day, such as between meetings, while waiting for someone, etc. These are good times to do things like check the email or revise your "to do" list.

Don't Procrastinate

If a task is genuinely urgent and important, get on with it. If, however, you find yourself making excuses about not doing something, ask yourself why.

You may be doubtful about whether you should be doing the task at all. Perhaps you're concerned about the ethics, or you don't think it's the best option. If so seek counsel, or talk it over with colleagues, family or friends to see if there is an alternative that may be better.

Stay Calm and Keep Things in Perspective

One of the most important things to remember is to stay calm. Feeling overwhelmed by too many tasks can be very stressful. Remember — the world will not end if you fail to finish all the items on your day's "to do" list. Going home or getting an early night so that you are better prepared to tackle things tomorrow may be a much better option than inflicting your body with a lot of unhealthy stress.

Taking a moment to pause, pray, and breathe deeply 10 times can do a lot to put your priorities into perspective. Often, after such a time-out we find our view changes quite substantially, and the things that bothered us aren't such big deals after all.

Plan for Success

What is the point of exercising proper time management without a clear plan to take us where we want to go?

- "By failing to prepare, you are preparing to fail." — Benjamin Franklin

- "He who every morning plans the transaction of the day and follows out that plan, carries a thread that will guide him through the maze of the most busy life. But where no plan is laid, where the disposal of time is surrendered merely to the chance of incidence, chaos will soon reign." — Victor Hugo

- "Lack of direction, not lack of time, is the problem. We all have twenty-four hour days." — Zig Ziglar

- "The best time to start was last year. Failing that, today will do." — Chris Guillebeau

- "Remember that time is money." — Benjamin Franklin

- "Yesterday is gone. Tomorrow has not yet come. We have only today. Let us begin." — Mother Teresa

Questions to Ask Yourself Before I Say 'Yes' to Anything

by Jess Ekstrom, CEO and Founder of HeadbandsOfHope.com, Speaker and Author.

It's okay to value your time and energy -- in fact, it's important.

When I started my business at 19 years old, I was so hungry to grow it that I said yes to just about any conference, any opportunity and anyone who would meet with me. Any idea I had was worth pursuing.

It worked out. I grew my business to be my full-time job out of college, and we're now six years strong. I credit a lot of the growth to just showing up and saying yes. Opportunities couldn't happen to me if I wasn't there for them. For years, I've been armed and ready to seize every glimmer of opportunity and give it my all.

But before long I found myself still saying yes -- without the same energy as I did before. I would agree to meet for coffee so someone could pick my brain about their business idea when I was slammed with work. By the time I was done, my energy for my goals and dreams was depleted.

Last last year, I started to realize that saying yes to everything was putting me on a path to burnout. With still so much I wanted to do with my business, burnout was not an option. At the age of 26, I felt I should be speeding up, not slowing down. But, ironically, I realized the key to speeding up in the areas that I wanted to grow was actually taking time to slow down.

I began to think of myself every morning as a full cup of water (or cup of coffee, usually). Each effort I made that day was a drip out of that glass. When the glass was empty, I had nothing left to give for that day. With each action I took, I could mentally see the glass getting lower.

I became more selective about where I put my time and energy. Just as I might work with an accountant on allocating my funds for different projects I want to pursue, I wanted to direct my energy where it was needed. I wanted my glass each day to go toward things that meant something to me, not just because I felt like I had to say yes.

The first step to doing less is being selective about what you choose to take on. With that, I ask myself ...

1. What purpose does this serve?

In years past, I was committing to things because I felt like I was supposed to or I didn't want to say no. But, now, I ask myself *what purpose does this serve?* Will this help any of my goals? Will I learn? Will this help something that's meaningful to me? Or, perhaps one of the most important questions I've started to ask myself, will this be fun?

If you ask yourself these questions and don't feel compelled by your answers, then it is probably a good indication that you should pass on moving forward. But, I also try to stay away from transactional opportunities that serve just my business or professional career. It's OK to do things just for fun if it is a good value for your time.

2. Why am I afraid to say 'no'?

One of the biggest reasons I would say yes to things I didn't want to do was because I didn't know how to say no. I felt like it was a slap in the face to the person who was asking and I never wanted to offend anyone. But, I've learned that I don't get offended when people say no to me. In fact, I'd rather someone be honest with me than say they'll do something and flake out later. So what was I so afraid of?

I got more comfortable with my response of thanking someone for thinking of me, passing this time, and wishing them the best. Everyone might have a different way they like to let people down gently. Find what works for you and stick with that.

3. What else could I be doing with this time?

I started public speaking professionally when I was 20 years old. It's one of my favorite things to do, but I realized it drains a lot of my cup between traveling, prepping for the opportunity, and then giving my all on stage for an hour and meet-and-greets afterwards. I typically speak professionally around 30-40 times a year, but I get dozens of emails every week asking me to speak for free at an event. While I really do love that someone wants to hear my message,

I've learned how much effort goes into every time I get on stage and the true value that I bring.

Speaking for free initially was a great way for me to learn and build my credibility, but I've come to a point where I'm confident enough in what I deliver to hold strong to that value and I also understand the toll it takes on my energy, because I'm giving it my all.

Give yourself permission to value your time, even if that means turning down unpaid work.

4. Can I delegate this?

As a college solo-entrepreneur, I didn't have money to hire anyone two years into my business. I did everything, and that was actually a really good thing because I learned about all sides of the business, even the parts (like numbers) that I was uncomfortable with. But, once the business started gaining momentum, I could afford to hire some helping hands.

But, I was stuck with the thought *why would I pay someone to do this when I could do it myself?* I was wrong for two reasons:

1. A lot of people out there are better than me at a lot of things.
2. Even if I could do it myself, it's not the best use of my time.

Now, my team is 10 times more productive than I ever was as a solo-entrepreneur. My assistant also helps as a gatekeeper to my schedule to make sure I'm not over-committing myself, which is super helpful.

5. What is stealing my energy?

The quickest way to drain your energy is by worrying about what other people are doing and going crazy over things you can't control. Beyond my physical time commitments, having fear of others, trying to measure up to others, worrying, and anxiety will drain energy and leave no room for anything creative or meaningful.

6. How do I refuel?

This might sound crazy, but this whole revelation I had about the benefits of doing less began in July when I found a dog on Craigslist and less than 24 hours later, he was mine. Suddenly, my life took a shift. I was taking Ollie (my dog) on multiple walks in the park every day. I was waking up and playing with him instead of checking my email. I was snuggling with him on the couch at night instead of scrolling through social media. Then when it was time to work, I was totally refueled and present. Ollie gave me an excuse to slow down, be playful, and break up the day with fresh air.

Removing yourself doesn't mean you're falling behind, it's actually refueling you to make better use of your time when you're ready to work. I now realize I'd rather have four hours of high quality work where I'm refreshed and creative than 10 hours of work where I'm going through the motions and can't think straight.

My team always laughs at me because whenever I have a long car ride, I always call them with tons of new ideas. When the juices are flowing, they always say Jess must be driving right now. And the reason I have ideas while I'm driving is because I'm removed. I'm not on my phone or on the internet or talking, I'm just quiet with my own thoughts. When that happens, it opens a space in my head that grows great ideas. But, most of the time, that space is too cluttered from being overly connected.

Being so hyper-connected all the time, it's so easy to see snapshots into everyone's lives and feel like you're falling behind or you should be doing more. Our social norm is that being busy is a badge of honor. But, this year, I'd like to debunk that myth because being busy all the time is an inefficient (and not very fun) way to accomplish your dreams.

I'm still a work in progress, but visualizing my energy as a tangible element has helped me understand how I work and has given me permission to do less, so I can do more in the areas that matter.

So where do you go from here? Identify areas that are draining to your energy (either change your approach, delegate or remove). Target the goals that matter to you. Leave room for what lifts your spirits (like a dog!). Take a deep breath and start there.

CHAPTER 7

Organization: Keys to Being Organized

Importance of Being Organized

Unorganized spaces tend to make people lose focus on the task at hand and often make them feel overwhelmed which results in less productivity and more stress, which can lead to numerous health problems. Various studies have shown an average person wastes up to an hour a day or 4.3 hours per week in search of misplaced items, which leads to frustration in addition to lost time. Being organized, or not, impacts all areas of your life: relationships, work, finances, health, and happiness.

Benefits of Being Organized

There are many benefits to being organized that lead to successful, happy, healthy lives:

- More focus on goals—When your time is not consumed with looking for misplaced items/ information, you have more time to focus on your goals.
- Manage time more effectively—When you are organized, you are able to manage time more effectively and work more efficiently which leads to more time for additional projects.
- Prioritize your tasks—When you are organized, you are able to prioritize your tasks to ensure each one is completed in a timely manner.

- Save money—The old saying, "Time is money" comes to mind. When you are organized your time is not wasted on searching for misplaced items/information. In addition, when you are organized you are not spending money on items you may already have but are unable to locate.
- Increase profitability—When the first four benefits listed above are happening, profits increase.
- Prepare for unexpected—When you have the day-to-day, regular business tasks organized and under control, the unexpected "emergencies" can be handled much more efficiently.
- Boost in energy—Clean, organized spaces energize you.
- Improve sleep habits—When you are organized, you have a plan/timeline to accomplish your tasks; therefore, you won't lie awake worrying about those tasks.
- Better overall health—Being organized reduces stress levels, which lessens your chances of ulcers, headaches, strokes, and hear attacks.
- Better balance in life—When you are organized, you use your time more efficiently, which opens up time for your family and friends and/or to participate in a hobby.

Tips to Getting Organized

It has been established that being organized is important, but how do you become organized? The most important item to have in order to begin your journey to organization is a calendar—and only one calendar. Choose your preference of electronic or paper calendar, but it is important to only have one. Trying to maintain more than one calendar is a waste of time, as you will be recording items twice—once on each calendar. This could also cause you to miss important deadlines and/or appointments, if you record them on one calendar and get interrupted or distracted before recording them on the other one. You could also double book yourself in this way.

After choosing and obtaining the calendar of your choice, you must make it a habit to record each and every appointment, task, deadline, etc. on this calendar. It is a good idea to enter reminders of upcoming events before the actual event, depending on preparations you need to make beforehand. For example, if you need to present a report or plan at a meeting on Friday, you should give yourself a reminder a few days before the actual meeting to ensure you have the report/plan ready ahead of time, so you are not scrambling the morning of the meeting. You should also color code entries on your calendar. For example, you could use green for business entries, pink for personal entries, etc.

Next, you need to organize your current work space. It is hard to stay organized and be productive, when your work space is cluttered and a mess. Organizing your work space does not have to happen all at once, depending on its current state. You want to decrease your stress and increase your productivity, not vice versa; therefore, it may be better to accomplish the organization of your work space in small chunks of time rather than all at once. The following are recommendations to unclutter your work area:

- Give yourself deadlines but be realistic. These deadlines will depend on the current state of your work area—how much time it will take to get it organized. For instance, you should not devote three solid days to organizing your work space and fall behind on all other tasks. Remember to record your deadlines/milestones on your calendar.
- Start by getting rid of anything you have not used in the past year. This includes old electronics, office supplies, etc. Either trash them or donate them.
- Create/update your filing system. Be sure to clearly label each file folder, be it electronic or paper. Be sure to back up all electronic files. In most cases, there is no need to have paper files at all, and if you can eliminate paper files by scanning and storing them electronically, this will be a huge step towards uncluttering your work area.
- Clean out desk drawers. Again, get rid of any items you have not used in the last year and scan any documents/paperwork you can to file electronically. It may be beneficial to obtain drawer organizers for the items you will keep in desk drawers. For convenience and efficiency place items used together in the same drawer, i.e., stamps with envelopes.
- Clean your desk top. Take everything off your desk top. This will force you to deal with each item before returning it, if it belongs there. When you place items back on your desk top, do so with convenience in mind: Items you use most should be within easy reach. Store rarely or less used items in drawers or elsewhere to maintain a clean, uncluttered desk top.
- If you need files/paperwork within easy reach on your desk top, use a file sorter. You want to stay away from stacks of paper or files on your desktop. Having them in a file sorter also keeps them within sight as a reminder to deal with them.
- Look around your work area. Are there items that should not be there? Old decorations, gifts, etc? It is okay to have a few of these items but not to the point of distraction. If any of these items are just collecting dust, get rid of them.

ORGANIZE

Now that your work area is clean and organized, you want to keep it that way. Staying organized is much easier than getting organized. The following are a few things you can do to maintain an organized life:

- Create a prioritized list of tasks for each day. Some people like to create this list each morning, but it may be best to create it at the end of the day when everything is fresh in your mind. Either way you will start your day with a list to keep you on track.

- The first 20 minutes of each day should be devoted to your email and any snail mail you may have received. Take action on each of these to keep your inbox organized and eliminate stacks of mail on your desk. You may want to create a folder (electronic and paper) for articles you want to read later.

- As you work on your prioritized list for the day, be sure to work on your top priorities first. It is easy to fall into the trap of finishing lower priority tasks first, because they are easier to complete, but this leads to falling behind on the most important ones.

- Create a folder labeled "Meetings" and keep it in the file sorter on your desk. It is a good idea to place notes, reports, etc. you need for upcoming meetings. If you can print items you need for the meeting, as soon as it is scheduled and place them in this file, you will always be prepared. Even if the meeting gets moved up, you will have no reason to stress about it.

- Once a project is completed, put all materials together and file them or put them away.

- At the end of each day, straighten your desk. You want to come in to a clean desk each morning.

- Be sure to file at least weekly. You should try to file items as you finish with them; however, this is not always possible. Create a folder labeled "Needs Filed" and place those items in it that will go in a paper file or need to be scanned to an electronic file. At least once a week, or anytime you have a few minutes, go to this folder and take care of these items.

- Create a tax box. You can use a medium-sized box or plastic tote or two small ones for this. If you choose one medium-sized container, you will need a divider to indicate personal and business. If you decide on two smaller containers, label one "Personal" and one "Business." These containers should be used to store all tax-related documents such as bank statements, receipts, charity donations, etc. This will make tax time less stressful and more efficient, whether you take the documents to a tax preparer or have them at your fingertips, if you prepare your own taxes.

- Make a list of goals you want to accomplish this year and the specific steps needed to reach them. Be realistic in your goals but push yourself at the same time. In other words your goals should be attainable but not ones you will achieve without any effort on your part. Place this list where you will see it throughout the day.

- Think through new purchases. Before you buy something you will have to store, clean, and care for, be sure it is something you will actually use. You want to avoid cluttering your work area after working hard to get it organized.

Organizing your Office and/or Home

Cleaning and organizing your home is the next step in organizing your life, improving your sleep habits, and decreasing overall health risks.

- Create a playlist of songs you enjoy and that energize you; label it "Organizing List" or something similar. This will motivate you to keep moving, while you are organizing and cleaning.

- Begin by inventorying items in your home room by room and determining which ones you do not need. If this seems overwhelming, place bins in the middle of the room you are cleaning and label each "Donate," "Sell," "Trash," and "Undecided," respectively. This allows you to toss items in the appropriate bins with minimum distraction.

- Go through items remaining after your inventory. Place these remaining items where they belong, if they have a place. If you are unsure about any of these items, place them in a container to deal with later. You do not want to waste time or get distracted here.

- Analyze each room, determining areas that need work such as storage drawers, hallways, corners, etc. and organize the ones that need it. You should not worry about perfect organization in drawers containing items such as undergarments and silverware that you will be in every day.

- Go through your wardrobe—anything you have not worn in a year, donate, sell, or trash. Be sure clothing items that should be on hangers are hung neatly, and clothing items that should be in drawers are folded and placed in their respective drawers.

- Be sure to label all boxes you use for storage and store similar items together.

Just like organizing your office as previously discussed, once you have completed the cleaning/uncluttering of your house, you want to ensure you keep it that way. However, you do not want to spend every free moment redoing the work, so a few tips to help you keep your home organized are listed below.

- Every time you leave a room that has items in it which do not belong there, take at least one with you and place it where it belongs.

- Spend 15 minutes (no more) picking up misplaced items each day. You do not want to have to clean every spare minute. If you follow these tips you will reach the point where you may not even need the 15 minutes.

- As previously mentioned in office organization, think through any new purchases. Will you really use it? Where are you going to store it? Can you replace an item you currently have with your new purchase?

- Use your calendar to keep track of personal appointments. Remember, you should only have one calendar, but everything should be on it to ensure no appointments are missed.

- Delegate responsibilities among family members. One person should not be responsible for all cleaning and picking up. At the very least, each individual should be responsible for ensuring his/her own items are placed where they should be.

- Try to wash, dry, and put away at least one load of laundry each day.

- Place a basket or any container of your choice by the front door to drop mail or other items you carry in the door with you and do not have time to deal with immediately. Be careful of this one—you do not want this "catch-all" to get out of control. Be sure you go through it at least once a week.

As mentioned previously, organization, or the lack thereof, effects our ability to be productive, and Cathy Sexton, in her article "7 Reasons Why Being Organized Boosts Productivity," reiterates this by stating, "...disorganization not only costs us time, it costs us money and for any business, large or small, the cost will have dire consequences on your bottom line." However, work is not the only aspect of our life effected by disorganization. Dr. Kathleen Hall, internationally recognized lifestyle expert in stress, work-life balance and mindful living states, "Being surrounded by clutter and disorganization causes stress. When you can't find your keys or a report you wrote, this causes you to release stress hormones. Producing more cortisol, a stress hormone, escalates the problem causing irritation, stress and memory loss." The American Heart Association's website states, "And your body's response to stress may be a headache, back strain, or stomach pains. Stress can also zap your energy, wreak havoc on your sleep and make you feel cranky, forgetful and out of control." Therefore, organization is important in and for all aspects of our lives.

CHAPTER 8

Change

Everything Changes

There is nothing more constant than change. We all experience change in our lives. It happens all around us, every day, all the time from the day we are conceived to the day we die. Given this fact, why do most people have a tendency to fear and resist change? One of the biggest reasons is we tend to get comfortable in the way we are doing things. No change is easy; therefore, most of us resist it. We like our world to remain constant. When something takes us out of our comfort zone—loss of job, change in a relationship, move to a new town, shock to our financial condition—we feel uncomfortable, because we have to make adjustments in unfamiliar territory.

However, there is always something valuable to be derived from difficulty, adversity or change; we just have to look for it with open minds. Winston Churchill once said, "A pessimist sees difficulty in every opportunity; an optimist sees opportunity in every difficulty."

When we accept the fact that nothing lasts forever, we will be better equipped for our journey. Struggle occurs when we resist reality. Our natural response to any kind of change in our lives may be to tense up on all levels: physical, mental, and emotional.

Therefore, we need to learn to anticipate change and view it as an opportunity, when it occurs.

Change can be a beautiful thing. It helps us grow as a person and can lead to a better situation.

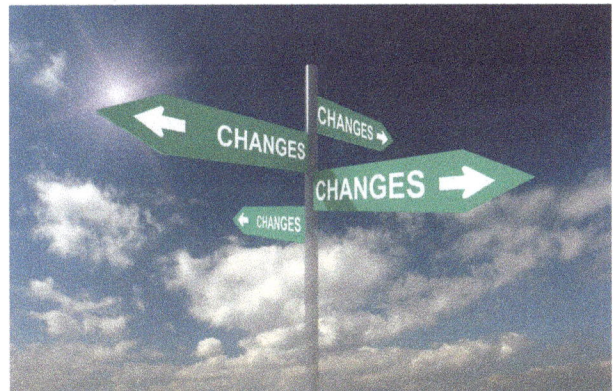

Why do some people easily embrace change, while others do everything in their power to prevent it? It all boils down to a matter of perception. We need to alter our perspective by think positively about the possibilities. When we acknowledge that there may be good ahead, we can focus our energy on productive actions that will carry us forward.

It is possible to go from resisting change to embracing change. So the next time you are faced with an experience which changes your reality, take a deep breath, take a step back and embrace it as an opportunity.

Methods for Embracing Change and Moving Forward

Change is difficult for people. For some, the difficulty stems from laziness, but for others the difficulty comes from a fear of a lack of control, uncertainty of outcomes, bruised egos, embarrassment or failing. People also seem to hate losing more than they love winning, and no one likes taking financial hits or enduring conflicts with others.

However, if we don't learn to embrace change and move forward we will not grow as a person, and we will be left behind. So, whether it's changing the focus of your business, having to learn new technology, replacing a prized employee, learning a new skill, or making adjustments in your personal life, it is important to know how to deal with change. We must be prepared to continually adapt for both survival and success. .Fortunately, these skills can be learned—and if you change your thinking, you can change your life. Here's how to get started.

1. Self-loathing is counterproductive to change.

Feeling sorry for yourself will often only lead down a negative path. It is counterproductive to change and will stagnate your efforts. Stay focused, set your sights high, and strive to achieve greatness. As you stay focused, change will becomes easier and easier. Wherever you are in life, maintain a positive attitude and remember it takes time to accomplish anything worth achieving.

2. Embrace change

The only thing constant in life is change. It is inevitable and no matter how happy we are currently, life and circumstances will always change. Embracing change is key to your success and happiness.

3. Be willing to go back in order to move forward.

Visualize trying to jump across a creek. You can't just jump standing from where you are. You have to physically move backward in order to give yourself the momentum needed to run and take that leap forward.

In the same way, when you face change, there will be a time when you will have to move backward, before you can move forward. That means reflecting on how you got into a situation, thinking about what you should have done differently, and learning from your mistakes.

4. Check your ego.

Typically, the biggest roadblock to change is you. Quiz yourself about the downside of pursuing change. What's the worst case scenario, if it doesn't work out? If the downside is primarily concern is a fear of failure or people laughing at you, it is time to get over it.

5. Take small action steps.

After you get your mind wrapped around the concept of embracing change, the first tweak is to take small steps forward. You are not going to be able to effect a wholesale change overnight, so find one small thing you can do then another and another and so on.

Start with an end goal and break it into small action steps. These small steps make change palatable and easier to accomplish.

6. Fail correctly.

Failure, when done properly, is a good thing. Taking on risk with change is part of the process and necessary.

However, the right way to fail means doing it quickly, inexpensively and never the same way twice. You don't want to have too much money or time hinging on any one outcome. If you do, then failure is bad, taking time and money away from other opportunities.

Testing your route on smaller scales in rapid succession allows for the risks to be lessened. So, when possible, try something that doesn't cost too much or take too much time. If it works, take the next step. If not, your failure isn't financially (or otherwise) devastating. And, of course, you need to learn from your failures so you don't repeat them in the future.

Creating Change and Creating Success

If you were to start your life over tomorrow, what would you do that you are not doing now? What would you stop doing that you are presently doing? What would you change?

If you are not excited about the day, when you wake up in the morning, then you are probably not satisfied with the life you're living. You may know you *want* to change your life, but you may not know *how* to change it. Or you may not be sure what needs to change; but you know something has to give. Your mind is flooded with questions: How did I get here? Is this all there is? Am I doing it right? These kinds of open-ended questions tend to overwhelm us and don't lead to any concrete plans for making positive change.

In the following three-part guide, we will dive into the kinds of questions you should be asking yourself and divulging what to do with the insight you gain by asking those questions. This analysis will open up endless opportunities and equip you with the knowledge needed to change your life for the better.

Part One: Proper Insight

The journey of healthy growth and positive change involves an honest self-evaluation and thorough analysis of one's circumstances. We must shine the light on the shadows and get a better grasp on what is holding us back in all areas of life. By answering the following questions, you can identify some key insights that will enable you to move past the things holding you back..

Why do you want to change your life?

Are you feeling trapped? Helpless? Overwhelmed? Underappreciated? Typically, when someone is looking to make a change, he or she is faced with many challenges. When faced with a challenge, dig deep and be honest with yourself as to what is inspiring you to make a change in your life.

What parts of your life do you want to change?

Are you unhappy with your job? Do you want to lose weight? Do you want to go back to school? Are you having trouble in your relationship? Are you having financial difficulties? Do you want to save more money? If you could think of three things to change in your life, what would they be? It is imperative that you be honest with yourself. Only you know in your heart what it is you want to change. It should also be your desire to change and not based upon someone else's insistence.

What are your passions?

Is there something you find yourself wishing you could do? Somewhere you desperately want to visit? Do you wish you had more time to perfect a certain skill set? Are you longing for a new friendship? How about a closer relationship with your kids? Is there a product, service, career, or business you long to pursue? Pinpoint the thing or things you find yourself desiring the most to change or accomplish.

How willing are you to make change?

As much as we want to do something new, the idea of life-changing experiences can be scary and can keep us from taking the first step towards change. Are you willing to relocate for a dream job? Are you willing to break a habit? Are you willing to put forth the effort, make the sacrifice, or pay the price? What about decluttering and minimizing your belongings? Try to picture your ideal situation and determine the

extent to which change is necessary to get there. Are you ready?

Change is uncomfortable. However, it provides us with a more fulfilling life and is often a necessary step to accomplishing our goals.

Who Are You?

After you have a better idea of what and why you need to make a change, you need to get down to some basics before beginning the change process. You need to identify your motivating forces. Finding a sense of purpose in your life can be the difference between reaching your goals and just going through the motions of everyday life as if you are stuck in a rut. Without clarity in your vision of the kind of life you want, how are you going to get there? It's like hopping in a car on the way to a place you've never been without a map or directions.

How many successful people do you know that just wing it every day? Successful people know who they are, where they want to end up in life, and have a plan geared toward success. B Clarifying who you are, what motivates you, and what you want will help you make the right changes for a better life.

Following are questions you should answer if you truly desire to make life altering changes. For this exercise to be effective, resulting in positive and lasting change, it is best to write down your answers.

1. What beliefs are keeping you stuck?
2. What beliefs have you come to uphold?
3. Where did these beliefs come from?
4. Why do you have these beliefs?

Answering these questions will help you gain a deeper insight into who you are as a person. There are two different types of belief schemas: the beliefs we have about ourselves and the beliefs we have about others and the world.

The beliefs we have about ourselves can propel us forward in life or hold us back. Saying things like "I'm a failure," or "I don't deserve happiness," become self-fulfilling prophecies and will keep you from being successful. One of the first and best things you can do is change your belief system. Positive self-affirmations, such as, "I'm worthy and capable of enjoying a good quality of life and success," or "Even if I have set-backs, I have the courage to try, and I am learning and growing as a person." A positive belief system and constructive thoughts leads to success.

Answer the following questions:

1. What beliefs do you have about yourself that limit you?
2. Why do you feel this way?
3. Does this reasoning make sense to you?
4. What do you want to accomplish?
5. Why haven't you been able to accomplish it before now?
6. What beliefs are stopping you?

Your Personal Mission Statement

Most organizations and businesses have a clear vision of why they do what they do and how they intend to do it. This vision, when clearly stated, is a mission statement. It captures an individual's, organization's and business' overall goals and summarizes them in a nice, neat statement. It is just as important for individuals to have a mission statement to guide them as they make plans and choices along their journey in life.

Creating a personal mission statement gives you clarity and provides a motto you can integrate into your day-to-day life. This personal constitution will motivate you and keep you focused.

Identify your values, your goals, and why things are important to you. Below are some sample mission statements to help get you started.

1. Through patience and effective communication (your value), I want to begin providing patient advocacy services to cancer patients (your goal) in an effort to ease the confusion and hardship that often accompanies a diagnosis (the why).

2. I am a strong-willed individual (your value) who envisions an alcohol free future (your goal) to improve my health, productivity and sense of well-being (the why).

3. As a driven and compassionate person (your value), I will begin attending nursing classes and move towards a new career in nursing (your goal) to increase my happiness, my salary and my purposefulness (the why).

Your mission statement should inspire you. Expect your mission statement to change throughout your lifetime as you change certain aspects of your life such as your career, interests, and skills. Write down your mission statement as a goal for your future. Each day going forward, ask yourself if what you are doing helps you reach your goal.

The additional questions below can help you gain a deeper understanding of yourself and how to make the right changes in your life.

- What are your current interests?
- What are some other interests you would like to try?
- What are you passionate about?
- What energizes you?
- What are your strengths? Focus on your strengths.
- What are your weaknesses?
- Are you reliable and responsible?
- What three words would your family and friends use to describe you?

You may be surprised when you re-read what you wrote as you discover new insights about yourself. You now have a good idea where you are and know what you need to work on.

As you reflect on your answers, think about making a vision board. In the same way writing helps us visualize a goal, crafting a collage of magazine clippings, printed material and other tidbits can be a daily motivator and a way to sort through your thoughts.

Overcoming Obstacles

There will always be lots of reasons why you can't, won't, or don't do something. Don't let transforming your life be held back by these negative words. You may feel scared, angry, or tired. You may also feel like you don't have enough time to get anything done. Obstacles like fear, poor self-image, low energy, discomfort, procrastination, perfectionism, lack of support, timing, no motivation, or uncertainty all play a part in your struggle to commit to change.

Fear is the first and most common obstacle people encounter when thinking about changing their lives. Fear of failure, fear of success, fear of embarrassment, and fear of the unknown can all stand in the way of making any changes, but only if you let them. Identify your fear and think about what you can learn by trying. Thomas Edison once said, "I have not failed, I have just found 10,000 ways that do not work." Without failure, we never learn how to better ourselves. Without making mistakes, we have nothing to learn from and cannot get any better.

Another source of fear can come from being afraid to step out of your comfort zone. The days you are most uncomfortable may be the days you learn the most about yourself. If you go to the same coffee shop, work the same job at the same desk, and go home to the same house thinking the same things, you will find very few opportunities to learn anything about yourself. You have to create the conditions in which change and growth are possible for you. Go to a different cafe, join a new club, or take a different route. Practice doing things that take you out of your comfort zone. The more you do outside your comfort zone, the easier it becomes, and the more you will grow as a person.

It is OK to feel the change-related distress and all of the associated emotions that come along with it. Process those emotions, put them in perspective, and keep moving forward towards your dreams.

Part Two: The Essential for Changing Your Life

"If you want something you've never had, you have to do something you've never done."

– Unknown

A Positive Attitude is Essential

If you are serious about creating change in your life, you must first change the way you think. A positive attitude is arguably the most important factor in your journey. It has the power to provide advantages in every area of your life. Think positive, and you will be much more likely to experience positive results.

Emotional Intelligence

Emotional intelligence (EQ) is a measurement of your ability to understand and take charge of your emotions, as well as your capacity to recognize the emotions of other people. You've probably heard of an IQ, which is based upon a measured score derived from one of several standardized tests designed to assess human intelligence. Interestingly, studies have shown that EQ is approximately twice as important as a predictor of job success as IQ.

Researchers recognize five main areas of EQ: self-awareness, self-regulation, motivation, empathy, and social skills. How you operate within these five areas dictates how effective your performance is at work, at school, and in life.

Self-awareness allows you to be more in tune with yourself, creates self-confidence and increases self-worth. It allows you to better forgive yourself for past errors and improve upon your weaknesses. It is also a critical component involved in creating healthy relationship boundaries with friends, family members, and co-workers. Furthermore, it enables us to better say "no" when necessary. Saying "yes" frequently to demands upon our time leads to burn out and resentfulness.

Self-regulation means exhibiting self-control, being honest with yourself and others, being conscientious, maintaining an open mind, and remaining flexible when an opportunity for change presents itself. With a high EQ on self-regulation we are able to maintain the discipline to complain less and focus more on finding solutions.

Motivation is what causes us to act or behave a particular way. Motivation gives us the drive to achieve, provides us the ability to commit to a goal, gives us the initiative to be ready when opportunity comes knocking, and helps us remain optimistic.

The empathy component of EQ is just as crucial as the other attributes listed above. Empathy is the ability to understand and share the feelings of another. To demonstrate empathy you should learn to recognize when people are happy, sad, frustrated, or confused and use this information to discern how you should interact with them. Sending the wrong signals to someone, because you incorrectly interpret them or their feelings typically causes unnecessary conflict.

EQ is something that can improve over time with dedication. The culmination of empathy, self-regulation, and self-awareness are improved by paying close attention to your thoughts throughout the day, during your day-to-day choices, and while interacting with others. It requires a commitment and intentional thinking to stop and absorb the moment you are experiencing, exercise self-awareness, practice self-regulation and strive to have empathy while listening to others.

Respond instead of React

Recall a situation in which you have felt you have no control. Perhaps it was when someone cut you off on the roadway. We've all been there. At that moment you have a choice, even if you don't feel like you do, as to how you will respond to the situation. We can either react (natural negative approach), or we can respond (a healthier positive and constructive approach).

Using the above road rage example, here are two possible reactions:

1. You call the person names, throw your hands up in the air, and stress yourself about how unfair it was that this happened to you.
2. You say a prayer for the person and yourself, brush it off, turn the radio up, sing-a-long,

and realize that nothing you do at this point will change what happened. You didn't have control over the situation, but you do have control over how you respond and how you feel about it in the end.

Reaction #1 above adds stress, is unhealthy, is a waste of time, and diminishes life. It describes someone who feels they "deserve" better treatment and expect fairness at every turn in life. This high standard will undoubtedly lead to feelings of disappointment and frustration during everyday interactions.

Reaction #2 is healthier and allows us to focus our attention on constructive thoughts that add to our quality of life. Instead of stressing out over what should have been, you employ a positive response something you can control, which rewards you in the end. Wrestling with what should have happened is never going to change the past. So instead of saying, "That person cut me off, now I'm going to be late," or "I am going to show him/her," respond in ways that are healthy, constructive, and pleasing to God. This type of response generates positive results.

Although it is hard to stop the instantaneous negative feelings that immediately come upon us when we feel we have been wronged, we can control how we respond and turn the negatives into positives. The goal now is to recognize when we feel wronged or upset and decide how we can best respond to the situation. Remember, you cannot control what happens to you, you can only control how you respond to it. This is where you have true power. A positive response to a bad situation can actually bring about positive results.

It is difficult to change the things we do, unless we see how they are negatively impacting us and those around us. The next time you are faced with a situation that frustrates you, upsets you, or makes you feel like you were treated unfairly, stop and ask yourself, "why do I feel this way and is this reaction helping me?" If the answer is no, then reevaluate and respond in a way that is positive and constructive.

Conflict Management

How often do you reflect on an argument and think to yourself that the argument wasn't even worth it. Some things are worth fighting for, and others just simply are not. Successfully eliminating battle baggage from your life is simply knowing the difference between what is worth the fight and what is not.

When faced with an argument, first express that you want to provide some sort of clarity into the disagreement as long as it continues to support a respectful dialogue. Alternatively, you should not react in an argument in ways that generate opposition and defensiveness. In other words, if you are trying to clarify a point so the other person can understand where you are coming from better, you should go for it. However, for most effective results this needs to be done in a calm and collected way. If you are finding yourself getting defensive and worked up over the subject, stop.

Whether you feel as if you are about to get into an argument with someone or you are actively in the midst of one, conduct a silent evaluation. Ask yourself the following questions: What are you hoping to get out of this disagreement? Is it a matter of pride or proving a point for the sake of proving it? Are you defending yourself or just arguing for the sake of argument?

Learning self-awareness is a weapon that can positively transform every interaction, intention, and outcome of our lives. Instead of spending enormous amounts of time arguing, realize the result will likely remain the same, because you see the situation from different perspectives. You can't change the way the other person processes information or how he/she views your body language or words, but you can change how you respond to the situation.

If there is no reasonable resolution to the situation, take a deep breathe and let it go. Agree to disagree. Instead of arguing spend your time doing something beneficial and constructive. Employing these new tactics will typically improve the relationship and your overall well-being.

Speak Your Truth

Have you ever experienced a time when you'thought something over and over again but never spoke it aloud? Most likely when you finally did tell someone, it brought you better clarity, felt more real, and gradually brought about positive changes.

Sometimes our truths aren't gleeful. Sometimes they are tragic, woeful, and all too painful to speak aloud. However, these are the pieces of ourselves that are the most important to share with others who are supportive and qualified to help us. Sharing things that are bothering us with trusted others helps us find peace or an answer we have been painstakingly seeking. Perhaps it is an answer to a question that has ultimately been holding you back from making progress. Vocalizing to a friend that you are terrified of working a dead-end job forever, when your heart lies somewhere else isn't going to make the life-change fairy appear. However, it will often free you of an underlying notion of discontent and allow friends and family to help steer you in directions that will solve your problems.

Expressing the truth can also take other forms. Instead of telling a white lie to a friend to avoid an uncomfortable conversation, tell them the truth. Not only will it generate a more authentic and deeper relationship, you will feel better about yourself, and it will typically help the other person in his/her development as a person and/or professional. Being real brings about genuine relationships, a more honest approach to your life, and prevents you from carrying a burden. Speaking the truth enables you to lead a fulfilled and authentic life.

Gratitude and Positive Thinking

How often do you express your gratitude? You may say thank you when someone holds the door for you at the store or when someone compliments you, but how often do you go out of your way to express gratitude for everything you are blessed with? Each one of us needs to take the time every day to give thanks for what we have, for what others do for us, and for how others enrich our life.

During a 2012 study at the University of Kentucky, 900 undergraduate student participants were asked to respond to a prompt, then they were either praised or scorned. Those that received praise experienced a decreased desire to seek revenge, and they were less likely to retaliate when faced with future negative feedback. Gratitude has been proven to reduce depression and stress, and increase happiness. Through happiness we will be able to view our lives in a better light, be grateful for the little things, make better choices, and ultimately change the course of our future. An attitude of gratitude tends to lead to positivity.

You are probably familiar with the glass-is-half-full adage which illustrates two archetypes of people: pessimists and optimists. The more we reinforce positive thinking, gratitude, and optimism, the more likely we are to look at the world through a less negative lens, which ultimately affects how we act, interact, and react.

Changing the destructive thoughts to positive statements, when you experience setbacks in life, is the central skill of optimism. To make this change, every time a less-than-desirable event happens, instead of focusing on what went wrong, focus on what you have learned and the good that can come from it. Some of life's best lessons are learned during the most difficult times. When viewed with a healthy and positive perspective changing times can evolve in wonderful opportunities, incredible personal growth and/or professional success.

Growing as a Person

Making a significant change in one area of your life typically impacts many other areas and causes changes in those areas as well. Below are keys to making a positive change and a smooth transition.

1. Clearly Identify Your Interests

What interests you? What are your passions? Identify your current interest and investigate new interests that will engage you. Step outside of your comfort zone. Surround yourself by those who share your interest to pave the way for new friendships, new skills, and a sense of community.

Part of the reason you are seeking a life change is because what you are doing isn't working. The goal is to make changes that are a better "fit" for you.

2. Hobby

A hobby can involve music, web design, DIY projects, photography, pottery, writing poetry, woodworking, gardening, etc., etc., etc. Any pleasurable activity you turn to when you need an outlet can be considered a hobby.

An enjoyable hobby can provide a healthy and constructive escape from normal or mundane routines. Furthermore, a hobby can also have a profound influence on how we relate to the world around us, can challenge us and can provide us a safe space to emote our authenticity.

If you feel more connected, more alive, and more relatable when you do a specific activity, you have found a good hobby that works for you. The next step is to make more time for your hobby, improve and share it and most importantly have fun with it. If you haven't found a good hobby yet then consider this your call-to-action. Explore new things and find what inspires you.

3. Get Physical

There is no better way to increase your mood, your self-respect, and your health than working out. If gyms aren't your thing, that's fine. Instead, try doing a simple physical activity like going up and down a long set of stairs, running, hiking, bike riding, playing a sport, gardening, participating in an exercise routine at home via your DVD player, joining an exercise boot camp class, etc. Engaging in a fitness program will improve your health, give you an outlet for your stress , and allow you the opportunity to meet new people.

Physical activity is a very positive and rewarding life choice. It can increase your self-image, make you healthier, improve your outlook and make you feel more accomplished.

4. Eat to Win

It's no secret that how we eat greatly influences our life. Of course, when it comes to dieting, many people think of their desire to lose weight. For most people losing weight also means shedding a poor self-image, lack of confidence, and harmful habits that can lead to other health problems. If you have considered weight loss as one of your "what-I-need-to-change" stressors then make a plan to start addressing it right away. Don't put it off.

Other diet-related changes to consider are increasing your water intake, improving your gut flora, eliminating unhealthy substances from your diet, and being mindful to avoid lactose, gluten, processed food, alcohol and sodium. It is also important to take a daily vitamin/mineral supplement made by a certified reputable company.

5. Taking Care of Yourself

You are important. Your life has meaning. You can also have a positive impact on many other people. Therefore, it is important to look after your own well-being. Caring for yourself comes in many different forms. One of the most important things you can do is get enough sleep.

How is your sleep schedule? Are you getting enough shut-eye? A well-rested mind is more inclined to have a positive outlook, feel better, handle stress better, and accomplish much.

How often do you take a break? Sometimes just taking a brief timeout can be the reset button you need to be more productive.

6. From Time Drains to Constructive Time

How often do you take a break? Sometimes just taking a brief timeout can be the reset button you need to be more productive.

The busier our lives become the more essential it is for us to take time out to process, meditate, and pray. To be most productive, happy, and at peace we need to put our phones down, turn off the TV, step away from the computer, and take a time out.

Our present-day information overload is draining. Social media, entertainment news, politics, polls, sports, advertisements, movies, video games, online surveys, and article after article can be overwhelming. Most of us spend way too much time doing these inane things versus building intentional blocks of meaningfulness. Limiting our digital intake, or taking a full digital detox, has tremendous benefits. Taking regular time outs is good for our physical and mental health and provides us opportunities to be creative and make plans. While there is no internet in the middle of the woods, there is a better connection to more important things such as our peace of mind, our creative side, our ability to process previous bombardments, and our future.

The ever-present digital world is not the only major time and energy drain: Negative or toxic relationships, consuming unhealthy substances, self-medicating, and clutter can also literally suck the life right out of you. Don't let your productivity get zapped by minutia, unhealthy input, and distractions. Weigh the pros and cons of every situation. If you are having trouble kicking some bad habits consider seeking help from a professional, friends and/or family.

Keep in mind that everything we do either produces a positive reward or a negative consequence. How we spend our time, and the choices we make will either be advantageous or detrimental. Sometimes these positive and negatives are subtle, while other times they are significant. Sometimes the positive reward or negative consequence is immediate, and in other instances it may be substantially longer (i.e. days, weeks, years) before we feel the effect. Nevertheless, there is always a ramification of the things we do and choices we make.

It is also essential to declutter your life emotionally and materialistically. Equally as important is keeping commitments you've made to yourself. Practice using your time more efficiently until it becomes second nature. Declutter your house or your workspace for positive results. Our minds tend to mirror our environment. If it's chaotic and messy, there is a high chance your thoughts will be as well. Minimize belongings, donate or sell unnecessary items, or at the very least reorganize. Productivity flourishes better with organization.

7. Making a Positive Difference in the World

When we do things that helps others and/or our planet, we feel better, increase our self-confidence and gain a deeper sense of self-worth. Consider volunteering for a local organization on a regular basis (i.e. quarterly, monthly, or weekly).

When you volunteer, you also gain a sense of community and empowerment. Participating with an aid organization can put your own life in perspective, provide opportunities to meet other caring people, and help you feel more grateful. Volunteer work is also a great resume builder and an excellent marketing ploy while serving a need and making a positive difference in the world. You might also find that volunteering may be the boost you need to start making other changes in your life.

8. Change Your Circumstances

Think for a moment about where you live, where you work, your budget, and your appearance. How does it make you feel? Are there things or one thing in your life that upsets or frustrates you? What would you like to change in your life? What would you like to do differently? What would you like to do that you are not presently doing?

How far are you willing to go to make positive changes in your life? Would you consider quitting your job to find a new one? Would you consider relocating to a new town? How about going back to school

for a career change? What about radically changing your budget to eliminate expenses, aggressively pay off debt, or save money; even if it means significant changes to your current lifestyle? What about giving your appearance a complete makeover (i.e. clothing, hairstyle, weight loss, etc.)? Would you be willing to put all of your stuff in storage for awhile in order to travel overseas to experience a different culture or go on a volunteer mission?

The thought of any of these options may immediately bring anxiety, fear, and a notion that "there's no way." However, getting out of your comfort zone and making a big life change can make a significant positive difference in your life. You might also discover your fear transitions into a sense of thrill, excitement, and eventually fulfillment.

Navigating these kind of extreme decisions will be challenging but don't let fear stop you. Listen to your intuition, establish your goals, plan your course of action, and then go for it. Use your present circumstances as a motivator and a starting point.

Part Three: Bringing it to Fruition

Hopefully the two above parts gave you some things to think about. Changing just for the sake of changing is rarely a prudent move. However, most of us can think of areas in our lives where change is needed. Once these areas are identified it is time to put things in motion. Following are some suggestions on how to change your life by making the above recommendations your reality.

Set Realistic Goals

Just because you feel energized to change your life don't expect it to happen overnight. Most often change takes time, patience, dedication, discipline, and effort. Set a realistic timeline for yourself, punctuated with relevant and attainable goals. Put your goals and timeline in writing. Break your goals in to smaller manageable action steps. Cross off the items as you progress towards the attainment of your goal. Doing this will positively reinforce the choices you've made, increase your confidence, and give you more momentum.

One of the fastest ways you can overwhelm yourself is setting one massive milestone without a clear manageable path to follow. If you are unsure if your goals and/or timeline are reasonable, reach out to people who can help you figure it out and assist you in making a plan. They may also be able to help you better define the goal and/or or assist you in adjusting your life so the journey is efficient and more manageable.

Enjoy the Process

While the process will be difficult try not to view it as a hardship. Enjoy the ride as you navigate your way over obstacles to a better life and success. Don't take yourself too seriously. Remember to laugh, love, and enjoy the present. Be patient when necessary and forgive when you feel wronged. Be thankful. Pray for wisdom, the needed resources, and the strength to carry on.

Support System

Deciding to make life changes is not something you should attempt to do alone. Everyone needs at least one person to bounce ideas off of and to share the ups and downs along the way. Evaluate the people in your life. Do they encourage you? Or, do they hold you back? Remove yourself from toxic relationships, retreat from those that bring you down, and limit your time with individuals who drain your positive energy.

Set standards and boundaries for the people in your life. Don't allow those who reinforce negative behavior to get into your personal circle of wellness. Seek out mentors and people who can help you reach your goals.

Remember Why it is Important to Change Your Life

Making positive changes will allow you to:

- Find your true life path and purpose.
- Identify and overcome emotional roadblocks.
- Set and reach your goals by taking intentional action.
- Increase your confidence and recognize your innate value.
- Boost personal, professional, and social growth.
- Enjoy a better quality of life.
- Live the life you always dreamed of!

"The journey of a thousand miles begins with one step."—Lao Tzu

Words of Wisdom for a Positive Change

1. "A year from now you will wish you had started today." -Karen Lamb
2. "It doesn't matter where you are, you are nowhere compared to where you can go." -Bob Proctor
3. "Man cannot discover new oceans unless he has the courage to lose sight of the shore." -Andre Gide
4. "You miss 100 percent of the shots you never take." -Wayne Gretzky
5. "Even if you stumble, you're still moving forward." Victor Kiam
6. "Lay a firm foundation with the bricks that others throw at you." -David Brinkley
7. "In a chronically leaking boat, energy devoted to changing vessels is more productive than energy devoted to patching leaks." -Warren Buffett
8. "Change is the law of life. And those who look only to the past or present are certain to miss the future."—John F. Kennedy
9. "Don't say you don't have enough time. You have exactly the same number of hours per day that were given to Helen Keller, Pasteur, Michelangelo, Mother Teresa, Leonardo da Vinci, Thomas Jefferson, and Albert Einstein." –Life's Little Instruction Book
10. "Someone was hurt before you, wronged before you, hungry before you, frightened before you, beaten before you, humiliated before you, raped before you... yet, someone survived... You can do anything you choose to do." –Maya Angelou
11. "Nobody can go back and start a new beginning, but anyone can start today and make a new ending." -Maria Robinson
12. "By changing nothing, nothing changes." -Tony Robbins
13. "Today is the first day of the rest of your life." -Anonymous
14. "All great changes are preceded by chaos." -Deepak Chopra
15. "You're braver than you believe, and stronger than you seem, and smarter than you think." –A.A. Milne
16. "You must do the thing you think you cannot do." -Eleanor Roosevelt
17. "Never, never, never, never give up." – Winston Churchill
18. "Courage doesn't always roar. Sometimes courage is the little voice at the end of the day that says I'll try again tomorrow." -Mary Anne Radmacher
19. "20 years from now you will be more disappointed by the things you didn't do than by the ones you did. So throw off the bowlines. Sail away from the safe harbor. Catch the trade winds in your sails. Explore. Dream. Discover." -Mark Twain
20. "One day your life will flash before your eyes. Make sure it's worth watching." –Unknown
21. "Getting over a painful experience is much like crossing monkey bars. You have to let go at some point in order to move forward." -C.S. Lewis
22. "Sometimes good things fall apart so better things can fall together." -Marilyn Monroe
23. "Whenever you find yourself on the side of the majority, it's time to pause and reflect." -Mark Twain

24. "If what you're doing is not your passion, you have nothing to lose."—Celestine Chua

25. "Use what talents you possess, the woods will be very silent if no birds sang there except those that sang best." -Henry van Dyke

26. "The best thing you can do is the right thing; the next best thing you can do is the wrong thing; the worst thing you can do is nothing." -Theodore Roosevelt

27. "Nothing diminishes anxiety faster than action." -Walter Anderson

28. "Live as if you were living for the second time and had acted as wrongly the first time as you are about to act now." -Viktor Frankl

29. "If you do what you've always done, you'll get what you've always gotten." -Tony Robbins

30. "Each person's task in life is to become an increasingly better person." -Leo Tolstoy

31. "All our dreams can come true—if we have the courage to pursue them." –Walt Disney

32. "Here is the test to find whether your mission on earth is finished. If you're alive, it isn't." -Richard Bach

33. "Your life does not get better by chance, it gets better by change." –Jim Rohn

34. "If today were the last day of my life, would I want to do what I am about to do today?" -Steve Jobs

35. "Fear, uncertainty and discomfort are your compasses toward growth."—Celestine Chua

36. "The greatest mistake you can make in life is to be continually fearing you will make one."

37. "To create more positive results in your life, replace 'if only' with 'next time.'"—Anonymous.

38. "As soon as anyone starts telling you to be 'realistic,' cross that person off your invitation list." –John Eliot

39. "I can accept failure, everyone fails at something. But I can't accept not trying." –Michael Jordan

40. "Believe you can and you're halfway there."— Theodore Roosevelt

Bottom line: The only thing that remains the same is 'change'. Regardless of where you are in life you can improve your situation. You can transition or proceed into those areas of life that will give you the most fulfillment. Identify your preferences, define your goals, implore positive thinking, pray without ceasing, work hard, and make the necessary sacrifices to bring your dreams to fruition. Love others and be thankful along the way.

CHAPTER 9

Communication: Being Effective

Effective communication is important in many aspects of both business and personal life: problem solving, conflict resolution, relationships, and interpersonal interactions.

Everyone communicates in one way or another, but very few people have mastered the skill of truly effective communication. Breakdowns in communication occur all too often and usually lead to a wide range of social problems, such as hurt feelings, missed business opportunities, anger, divorce, and even violence.

Communication is both an expressive (message-sending) and receptive (message-receiving) process. Failure to communicate effectively can stem from problems on either or both ends of the process.

Good communication skills are key to success in life, work and relationships. Without effective communication, a message can lead to error, misunderstanding, frustration, or even disaster when misinterpreted or poorly delivered.

Communication is the process by which individuals or groups of people exchange information. In this process we try to convey our thoughts, intentions and objectives as clearly and accurately as possible.

Statistics point to the fact that approximately 85% percent of our success in life is directly attributable to our communication skills. That means that no matter how ambitious, how committed, or how highly educated someone is, he/she still has a low probability of success without the development of the right communication skills.

Communication is successful only when both the sender and receiver take away the same information from the exchange. In today's highly informational and technological environment, it is increasingly important to have good communication skills.

The inability to communicate effectively will hold a person back not only in their careers, but also in their social and personal relationships. The good news

is that effective communication is a skill anyone can learn. Although all new skills take time to refine, with effort and practice you can develop good, even exceptional, communication skills.

When you take the time to acquire and hone good communication skills, you open yourself up to better relationships, more career opportunities, and increased self-confidence. Moreover, you reach higher levels of mutual understanding and cooperation with others while successfully attaining your goals.

These Key Components Lead To Effective Communication

Every human communication interaction, whether face-to-face, written, via telephone or other means, has three critical components:
- Sending Communication
- Receiving Communication
- Feedback.

Sending Communication

The first component of communication is sending communications. Communication scholars refer to this as "encoding" a message; otherwise referred to as constructing and transmitting an understandable message to the receiver.

Think Results. As you create the message you need to transmit, ask yourself how this communication could potentially drive overall company goals. For example, you might need to address a particular employee on how to improve his customer service. Rather than focus the communication on failures, effective communicators focus the conversation on results. For example, remind the employee how his actions drive company goals and impact overall results.

Clear Purpose. Every message you send should have a clear purpose. For example, your outcome could be reaching a specific goal, solving a customer issue, giving information, seeking information, building

a relationship, etc. Whatever the purpose of your communication, stick to it.

Think Before You Speak. This may come across as elementary advice, but it is amazing how many well-meaning people engage their mouths before their brains. If you prefer to shoot from the hip and just say what's on your mind without thinking, you will open yourself (and your company) up to potentially damaging lawsuits and employee turnover. At the very least you will eventually come across as being foolish or untrustworthy.

Structure the Message for the Receiver. Since mutual understanding is our goal, to be an effective communicator you must create messages that resonate with their receivers (employees, colleagues, customers, suppliers, and personal acquaintances). Too often we spend more time thinking about what we want to say rather than what others need to hear. That is the difference.

Avoid Irrelevant Details. Do you know someone who, when telling a story or relating some information, includes so much irrelevant detail that you want to scream? To send effective messages, it is important to focus only on the relevant information, data or contexts.

Read Reactions and Adjust. Baseball Hall of Famer Yogi Berra is often credited for saying, "You can observe a lot just by watching." The same is true in communication. As you create and transmit your communication, watch for the reactions of others. Be prepared to read the reactions and adjust future communications in order to achieve intended results.

Focus on Results. This key point is similar to a point mentioned above, but it is worth repeating, since it is the most important component of sending messages. Remember to always keep the results, the ultimate outcome of the communication, in your mind as you communicate with others.

Receiving Communication

Most people believe they are great listeners, yet studies have shown that most people are only average, or more realistically, poor listeners. Without strong listening skills, growth opportunities are limited. Listening needs to be an active-process as opposed to a passive one.

Active Listening Tips

- **Listen with purpose.** Ask yourself "What worthwhile idea is being expressed?"... " What is being said that I can use?"
- **Judge content not delivery.** Look beyond the speaker's delivery and concentrate on what is being said.
- **Keep your emotions in check.** Avoid becoming over stimulated by what the speaker says. Avoid allowing your own biases or values to detract from the speaker's message.
- **Listen for the main idea or central themes of the message**.
- **Be flexible.** Find a variety of ways to remember what you hear. Find a variety of note keeping techniques to help you remember.
- **Work at listening.** Give your full attention to the speaker. Face the speaker. Use facial expressions that indicate you are following what the speaker is saying.
- **Resist distractions.** Concentrate on what the speaker is saying. Stay focused on the situation at hand.
- **Keep an open mind.** Avoid developing blind spots regarding cherished convictions. When you hear "red-flag" words keep your emotions in check.

Examples of Active Listening		
TYPE OF STATEMENT	DESIRED RESULTS	EXAMPLE
Encouraging	Convey interest Speaker continues to talk.	"I see…" "That's interesting…"
Clarifying or open-ended questions	Obtain more information	"Can you tell me more?" "Is there anything else?" "How do you see the situation?"
Restating	Show speaker that Listener is listening Let speaker know Listener understands facts	"As I understand it, your idea is… "Do you mean, for example…" "In other words, this is what happened…"
Reflecting	Shows understanding of speakers view of situation	"You feel that…" "You seem pretty concerned…" "Sounds like you…" "So you would like to see…"
Silence or Pause	Encourages reflection. Allows speaker to fully express ideas.	
Summarizing	Shows grasp of situation or problem Highlight key facts and ideas.	"These seem to be the key ideas you have expressed…" "Let me summarize…"

- **Capitalize on thought speed.** Most of us talk about 120 words a minute. Our thinking speed is about 500 words a minute. Thus there is a lot of time to spare while a person is speaking to us. Don't let your mind wander and then dart back to the conversation. Put the spare time to good use by thinking about what is being said; trying to anticipate the point; mentally summarizing the point so far; mentally questioning any supporting points; looking for nonverbal clues to the meaning.

Monitoring Your Listening Skills

Poor listening skills account for the majority of communication problems. Listening is how we find out

Types of Listeners	
Detached • Avoids making eye contact • Appears withdrawn • Lacks enthusiasm • Seems inattentive, disinterested or bored	**Passive** • May or may not make eye contact • Fakes attention • Uses little energy or effort • Appears calm and laid back
Involved • Provides some direct eye contact • Has an alert posture • Gives the speaker some attention • Reflects on the message to a degree	**Active** • Has an alert posture • Uses direct eye contact • Gives full attention • Focuses on what is said • Participates fully

people's preferences, desires, wants, and needs. It is how we learn to customize our message to others.

Good listening is not just looking at someone and nodding your head in agreement. You have to acknowledge what is being said and let the other person know that you understand. The more you can acknowledge what is being said, the greater ability you have to persuade and influence.

Why is listening so difficult for most of us? Why is it that when two people get together and talk, they both walk away with two completely different views about the conversation?

Top Five Challenges to Listening Effectively

1. **Thinking about our response instead of truly listening and thinking about what the other person is saying.** When we are mentally planning our own agenda and game plan instead of what the other person is trying to communicate, it diminishes the desired result. Even if we patiently wait our turn to talk-if we are not

actively listening—we will not obtain optimal benefits from the interaction.

2. **Not Concentrating.** We talk at a rate of 120 to 150 words per minute, but we can think 400 to 800 words per minute. This allows us time to think in between words being said. We can pretend to listen while really thinking of something else.

3. **Jumping to conclusions.** Sometimes we assume we know exactly what the other person is going to say next, and we begin forming reactions based on those assumptions. We start putting words into the other speaker's mouth, because we are so sure we know what they are talking about.

4. **Prejudging speakers on their delivery and personal appearance.** We often judge people by the way they look or speak instead of listening to what they say. Some people are so put off by personal appearance, regional accents, speech defects, and mannerisms that they don't even try to listen to the message.

5. **Lack of active listening training.** Some people don't know how to listen effectively. If they

haven't had any training or guidance on how to listen effectively, they may not be accustomed to or even realize the mental effort and level of involvement really required to do so.

Listed below are keys for effective listening. Following these guidelines will help you achieve optimal success.

1. **Give them your undivided attention.** They are the most important people in the world to you at this time — make them feel that way. Don't get distracted by your surroundings. Stop talking and concentrate on them.

2. **Look them directly in the eye while they are talking.** Lean forward to indicate interest and concern. Listen calmly and patiently.

3. **Show sincere interest in the other person.** Simply things like not talking, nodding your head, and agreeing with verbal sounds like "uh huh" shows the other person we are interested in what they are saying. Listen for main points and don't interrupt.

4. **Keep the conversation going by asking questions.** Prompt more information from them by repeating their phrases and asking questions.

5. **Use silence to encourage the other person to talk**. As the old adage goes "silence is golden." Being silent encourages others to talk more, which often results in a wealth of valuable information..

6. **Pause before replying or continuing.** Wait three to five seconds and reply thoughtfully. Don't leap in, even if you know the answer. When you pause, it shows the other person you consider what they are saying is valuable. If you apply your listening skills, you will be able to glean golden nuggets of information from your audience. Since truly effective communication means adapting your message to the person you are talking to, listening is crucial. Pausing for silence also shows you

Active Listening Skills

Eye Contact
Eye contact during the conversation shows the speaker that you are giving them your attention and that you really care about what they say.

Avoid Distractions
There are many examples of distractions such as our thoughts, mobile phones, gadgets, music, side activities, other people, and more. Learn to avoid these distractions otherwise they can destroy conversation.

Body Gestures
Body gestures and language are a whole science. Your body gesture tells the speaker whether you listen carefully or not.

Give Feedback
Ask questions to clarify certain points, tell your opinions, summarize the speaker's comments.

Show That You're Listening
Use facial expressions such as a smile, note your posture, encourage the speaker to share and to continue.

Listening allows you to learn, have relationships, to plan, to develop, to be part of something, to create, to think, and much more.

are interested in your audience and stimulates interest in the conversation

Feedback

After sending the message and receiving a response, it is time to offer feedback to complete the communication process. Interestingly, feedback has two parts—giving and receiving. Effective communicators often give feedback to employees, colleagues, vendors, customers and even bosses. Other times effective communicators receive feedback from the same groups. Following is an examination of each part separately.

When giving feedback, we need to consider these eight points:

1. **Positive.** Studies indicate that we hear "No" 4-7 times more than we hear "Yes." Rewording, rephrasing, and restructuring our feedback in a way that produces a 3:1 ratio of positive-to-negative feedback produces much more favorable results.

2. **Constructive.** Destructive, mean-spirited, intimidating, or demeaning feedback is ruinous to relationships and destroys morale, team spirit and productivity. Moreover, it impairs your ability to effectively lead your team to achieve great results. In all cases, phrase your feedback so as to help the person—never to tear them down.

3. **Focus on behavior—not personality.** Even though you may think the other person is an unmitigated jerk, feedback should focus on their behaviors. People tend to have much more control over their behaviors than they do their personality.

4. **Non-Judgmental.** Suspend your internal judgments and focus only on the issue at hand. For example, you may not approve of a particular employee's hairdo, choice of jewelry, or style of music. These issues, and dozens more like them, seldom have any direct consequence on performance, and they should not have any direct

impact on your feedback as long as they are not negatively impacting the employee's performance at work.

5. **Clear.** Effective communicators have a crystal clear picture of what they are trying to accomplish with the feedback. Make sure you are clear on what you desire as the outcome of the feedback and be prepared to provide examples and backup.

6. **Concise.** Avoid needless rambling, injecting irrelevant information, or "dancing around" the issue. Be professional, courteous and concise in stating the feedback. Get to the point.

7. **Specific.** Avoid generalities like, "Jim, you need to do a better job." Ask yourself what, specifically, Jim needs to improve. The more specific the feedback, supported with quality measures, the more effective the communication.

8. **Follow with Appreciation.** First, saying "thank you" expresses your sincere appreciation (without sounding condescending). Second, it personalizes the relationship between you and the other person. Third, it demonstrates your honest, non-manipulative interest in improving the relationship and/or the other person's performance.

Similarly, think of these six points when receiving feedback:

1. **Be open-minded.** Even when you disagree, or the feedback seems inaccurate, keep your mind open and your mouth shut. More often than not, the feedback you receive will help you grow as a person.

2. **Consider all feedback as constructive.** Effective communicators assume all feedback (be it from employees, bosses, customers, or spouses) contains something they can utilize to improve themselves. Approach all feedback with the mindset that what you are about to receive is constructive, not destructive.

Avoid Communication Stoppers

Behaviors and phrases that often stop a person from communicating:

⚐	ORDERING	"Don't talk like that."
⚐	WARNING	"If you do that, you'll be sorry."
⚐	MORALIZING	"You ought/should...."
⚐	ADVISING	"I suggest that you..."
⚐	REASON WITH	"Let's look at the facts."
⚐	DIAGNOSING	"You feel that way because..."
⚐	JUDGING	"You are wrong about that."
⚐	NAME CALLING	"You are acting like...."
⚐	DISTRACTING	"Let's talk about something else."
⚐	INTERRUPTING	"But what about..."

3. **Listen non-defensively.** Effective communicators realize that listening non-defensively will achieve the best results.

4. **Restrain your tongue.** It is natural and easy to immediately lash back at those who offer us well-meaning feedback. However, this is a mistake. It is not productive to the conversation, hurts relationships, and inhibits growth. Let others speak without interrupting or becoming defensive.

5. **Try not to take comments personally.** When you receive feedback, maintain a thick skin and do not assume you are being attacked as a person, instead think of it in terms of behavior: How can you use the feedback to grow and improve.

6. **Say thank you.** It takes valuable time, energy, and courage for a person to offer you feedback. Be thankful for it, even when it hurts or does not add a lot of value to end results. By graciously accepting the feedback, you communicate your willingness to listen, grow and have a better relationship with the other person.

Communication is an active, two-way communication process. To achieve optimal results and become an effective communicator, it is important to master the three essential components of communication discussed above: (1) sending communications, (2) receiving communications and (2) feedback.

Communication Improvement Tips

- Open the door to two-way conversation. Encourage questions. Ask questions.

- Consider the sender/receiver's communications strengths and weaknesses, and communicate in the manner best accepted by the sender/receiver.

- Paraphrase what you heard back to the speaker to ensure you have a common understanding.

- Don't be thrown off course by words that affect you emotionally.

- Continue to listen, even when the urge is to start a debate.

- Communicate to be understood—not to impress

- Use short words that communicate clearly and concretely; present one idea, at the most two ideas, in one sentence.

- Avoid jargon. Use strong verbs. Avoid passive voice as much as possible.

- Be open to feedback.

- Be an active listener.

Non-Verbal Communication Tips

65% of the message is sent non-verbally. Become sensitive to non-verbal messages. Look for such things as body position and movement, gestures,

facial expressions, eye contact, silence, use of space and time, etc.

- **Eye Contact.** If you look someone in the eye, they pay more attention to what is being said.
- **Posture.** Good posture is a sign of confidence and creates a sense of trust in your skills and abilities.
- **Gestures.** Use only the body movements and gestures necessary to make your point. Excessive motions are distracting to the listener.
- **Expression.** Your expressions sometimes say more than the words you speak. Try to smile and demonstrate interest in the topic.
- **Voice.** Speak with a firm and assertive voice. This tone implies directness and honesty.

Negotiation: Keys to Success

Communication is a key component to favorable negotiation. There are specific characteristics and attributes which contribute to successful negotiations, and these skills are worth developing, if you hope to improve your situation and life.

Negotiation is something everybody does whether they realize it or not. It's something we do in our work as well as in our personal life. For example, we negotiate on many work-related issues with co-workers and supervisors regarding schedules, resources, staffing, and timelines. We negotiate with our significant other over everything from chores to entertainment options. We regularly negotiate over many issues with various people in our community at some level or another.

The fact is that whether we are aware of it or not, we spend a significant part of our day negotiating. That is why good negotiation skills are so important. Studies show that having good negotiation skills plays an important role in business success and in our quality of life.

Being a good negotiator allows us to build, maintain, and improve important professional and personal relationships which leads to becoming more respected. Additionally, being a good negotiator makes us a more efficient person. Instead of spending hours arguing with people or trying to force them to do what you want, you can reach agreements, find solutions to tough problems, and keep work moving ahead more easily and with less effort, if you negotiate effectively. Finally, being a good negotiator helps you achieve important goals and get what you need and want for yourself and your organization.

Following are the key ingredients to becoming a good negotiator. The more you develop these skills, the better negotiator you will be.

- **Integrity** includes honesty and trustworthiness — qualities necessary for others to trust you in a negotiation. You can hone your integrity by following rules, keeping agreements, following through with what you say you will do, and being steadfastly honest. Not replying to a question is not acceptable. Blowing off a commitment, divulging restricted information, and lying are not acceptable.
- **Empathy** is the ability to participate in another's feelings or ideas: putting yourself in someone else's shoes. It's the bedrock of successful communication and a necessary trait for great negotiators. Being empathetic helps you recognize the differences between you and other people and allows you to maintain your identity as you experience the views and emotions of others. You can improve your empathetic responses by acknowledging the behaviors, values, and goals of *others* and taking time to consider things from other people's perspectives.
- **Self-discipline** is the ability to complete tasks and reach goals without someone else directing or motivating you. Self-discipline is an internal force that drives us toward our goals. It motivates us to keep moving forward despite challenges,

nay-sayers, and other persuasive external forces pushing against us.

- **Stamina** is the ability to keep going when others drop by the wayside. Stamina is an attribute of all great negotiators. After all, you can't win the game, if you don't have the stamina to keep playing and doing what is necessary to achieve success.

- **Flexibility** is the ability to deal with new situations and unexpected obstacles. Flexibility is at the heart of closing a deal in a way that satisfies each side and works in the real world. If one approach doesn't work, try another.

- **Patience** is the act of bearing pains or trials without complaint. This includes tolerating frustration and adversity on the way to reaching your goals without giving up. All successful people know that being refused, delayed, and blocked is part of life. Success comes to those who are steadfast and patient.

- **Respect** tends to be reciprocal in negotiations: If you give it, you're much more likely to get it back. In order to respect other people and the limits they set, you must first have respect for yourself and the limits you set.

- **Fairness** is the state, condition, or quality of being fair, or free from bias, injustice, and discrimination. To develop fairness, consider your goals and those of the other party. Identify areas of agreement and areas that need compromise. Being fair ensures that a deal closes or ends with satisfaction for all parties.

- **Responsibility** is the ability to be reliable, be accountable, and accept consequences. Being responsible doesn't mean you won't make mistakes, but it does mean you will take the proper action to correct your mistakes when possible. You can improve your skills in this area by owning up to your mistakes and taking the appropriate steps to properly address problems.

- **Sense of humor** is the ability to see, appreciate, or express amusing aspects of various situations.

Finding humor in adversity helps you get on with finding solutions rather than getting stuck in the blame game. When you make a mistake or have a disappointment, laugh at yourself, and think of how to turn it into a story later. To see the humor in a situation, you need to take a step back. Developing a sense of humor helps you develop resilience and can ease you through your darkest times.

Public Speaking

Standing up in front of a roomful of people terrifies many people, even if those people are peers. Fear of public speaking is the most common of all phobias. It is a form of performance anxiety in which a person becomes very concerned he or she will look visibly anxious or possibly even have a panic attack while speaking. Over time people try to protect themselves by either avoiding public speaking or by struggling against speech anxiety.

However, resisting public speaking engagements will substantially reduce one's ability to capitalize on opportunities and limit their growth as a person. Being a bad public speaker can have just as many disadvantages.

Even if you don't need to make regular presentations in front of a group, there are plenty of situations where good public speaking skills can help you advance your career and create opportunities. Regardless of one's anxiety or poor public speaking skills we all have to speak in public from time to time, whether it is talking in a team meeting or having to address a large audience. For example, you might have to talk about your organization at a conference, make a speech after accepting an award, or teach a class of new recruits. Speaking to an audience also includes online presentations or talks which may include training a virtual team, or speaking to a group of customers in an online meeting. Good public speaking skills are important in other areas of your life as well. You might be asked to make a speech

Rhetorical Triangle

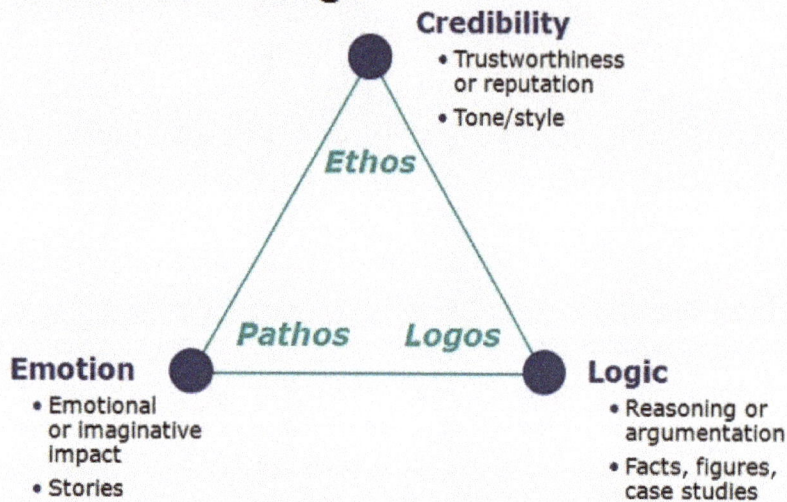

Credibility
- Trustworthiness or reputation
- Tone/style

Ethos

Pathos   *Logos*

Emotion
- Emotional or imaginative impact
- Stories

Logic
- Reasoning or argumentation
- Facts, figures, case studies

at a friend's wedding, give a eulogy for a loved one, or inspire a group of volunteers at a charity event.

Whether avoided, done well or done poorly, the outcome of our public speaking strongly affects the way people think about us. Good public speaking skills will open doors, and poor public speaking skills will close them. Being a good public speaker can enhance your reputation, boost your confidence, open up countless opportunities, and generate tremendous rewards.

The good news is that through proper preparation and practice, anyone can not only become a good public speaker but can perform exceptionally well. Following are methods to help you overcome fear and greatly improve your speaking skills.

Speech Preparation

The best thing one can do to have a successful speech is to plan and prepare appropriately. Using tools like the Rhetorical Triangle, Monroe's Motivated Sequence, and the 7Cs of Communication in your preparation helps lay the foundation for a successful speech.

A strong and powerful opening will grab the audience's attention. For example, you could start with an interesting statistic, headline, or fact that

pertains to your topic and resonates with your audience.

You can also use story telling as a powerful opener. Stories can change the way we think, act, and feel. They can form the foundations of an entire workplace culture, and they have the power to break down barriers and turn bad situations around. Stories can capture our imaginations, illustrate our ideas, arouse our passions, and inspire us in ways that cold, hard facts often can't. Stories can be powerful business tools, and successful leaders use them to engage their teams. Telling a good story can produce wonderful results.

It is important to keep in mind that occasions arise when you need to speak in public without being scheduled ahead of time. You can make good impromptu speeches by having ideas and mini-speeches pre-prepared. It also helps to have a good, thorough understanding of what is going on in areas in which you may be asked to speak, i.e. your organization, industry, etc.

Practice, Practice Practice

A great way to ensure your speech goes smoothly is to rehearse what you're going to say. There is a

7 Cs On Communication

1. Completeness
2. Conciseness
3. Clarity
4. Correctness
5. Consideration
6. Courtesy
7. Concreteness

good reason we say, "Practice makes perfect!" You simply cannot be a confident, compelling speaker without practice.

To get practice, seek opportunities to speak in front of others. Toastmasters International is a club geared specifically towards aspiring speakers and offers many speaking opportunities at Toastmasters sessions surrounded by encouraging people. You can also put yourself in situations that require public speaking such as cross-training a group from another department or volunteering to speak at team meetings.

You will also find public speaking classes at local colleges and learning centers. You'll not only learn great techniques, you'll be able to practice your skills in front of a roomful of people who share similar apprehensions about public speaking.

Practice alone often and make adjustments as necessary, until your speech flows smoothly and easily. Then, if appropriate, do a dummy run in front of a small audience. This will help calm your jitters and make you feel more comfortable with the material. Your audience can also give you useful feedback on both your material and your performance.

Use a timer to make sure your speech isn't too long and drawn-out or too short, missing key points. Another method used in the path to perfection is to videotape yourself giving the speech; this allows you to critique your mannerisms and plan out your pace. Practice will reduce anxiety and make sure your speech is successful.

Connect with the Audience

In addition to using an outline to guide your speech, make eye contact with people in the crowd to encourage engagement and read their interest. If the audience appears to be bored, infuse a spark in the speech through voice inflexion (putting a stronger (louder) emphasis on key points), and/or asking the audience a question leading them to your next bullet point.

Interaction with the Audience

When you speak, try to engage your audience. This makes you feel less isolated as a speaker and keeps everyone involved with your message. If appropriate, ask leading questions targeted to individuals or groups and encourage people to participate and ask questions.

Monroe's Motivated Sequence

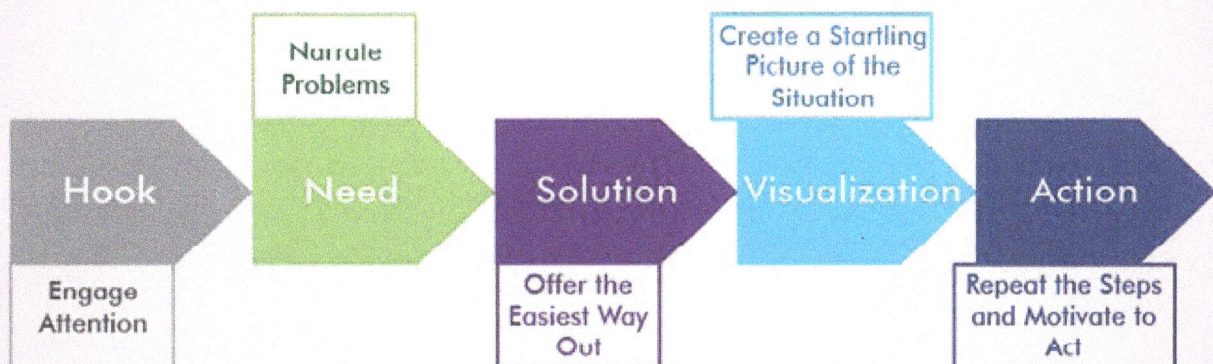

Hook	Need	Solution	Visualization	Action
	Narrate Problems		Create a Startling Picture of the Situation	
Engage Attention		Offer the Easiest Way Out		Repeat the Steps and Motivate to Act

Keep in mind that some words reduce your power as a speaker. For instance, think about how these sentences sound: "I just want to add that I think we can meet these goals" or "I just think this plan is a good one." The words "just" and "I think" limit your authority and conviction. Don't use them.

A similar word is "actually," as in, "Actually, I'd like to add that we were under budget last quarter." When you use "actually," it conveys a sense of submissiveness or even surprise. Instead, say what things are: "We were under budget last quarter" is clear and direct.

Also, pay attention to how you're speaking. If you're nervous, you may talk quickly. This increases the chances of tripping over your words or saying something you don't mean. Force yourself to slow down by breathing deeply. Don't be afraid to gather your thoughts; pauses are an important part of conversation, and they make you sound confident, natural, and authentic.

Finally, avoid reading word-for-word from your notes. Instead, make a list of important points on cue cards, or, as you get better at public speaking, try to memorize what you're going to say—you can still refer back to your cue cards as needed.

Body Language

Your body language will give your audience constant, subtle clues regarding your inner state. The audience will know if you are overly nervous, or if you don't believe in what you are saying.

Pay attention to your body language: stand up straight, take deep breaths, look people in the eye, and smile. Don't lean on one leg or use gestures that feel unnatural.

Many people prefer to speak behind a lectern when giving presentations. While lecterns can be useful for holding notes, they also put a barrier between you and the audience and have the potential of becoming a "crutch." The audience may find this undesirable.

Instead of standing behind a lectern, walk around and use gestures to engage the audience. This movement and energy will also come through in your voice, making your speech more inviting and engaging.

Think Positively

Fear makes it all too easy to slip into a cycle of negative self-talk. Before you speak, sabotaging thoughts such as "I'll never be good at this!" or "I'm going to fall flat on my face!" lower your confidence and set you up for a self-fulfilling prophecy.

Positive thinking can make a huge difference to the success of your communication, because it helps you feel more confident. Use affirmations and visualization to raise your confidence. This is especially important before you give a speech. Visualize giving a successful presentation and imagine how you will feel when it is over and you have made a positive difference for others or accomplished your goal. Use positive affirmations such as "I'm grateful I have the opportunity to help my audience" or "I'm going to do well!"

A Conversational Speech

As tempting as it may be to type up a speech and read it word for word, refrain from doing so. If you change your natural speech patterns to give a speech written like an essay, you are setting yourself up to fail. Audiences listen better, when the speaker talks to them instead of reads to them.

Build your speech around a structure of key points you want the audience to come away with. Structuring a speech in a conversational style will boost confidence, prevent you from sounding robotic, and help the audience perceive it as inviting.

Cope With Nerves

Many people cite speaking to an audience as their biggest fear. Public speaking can kick in your "fight or flight" response; adrenaline courses through your bloodstream; your heart rate increases; you sweat; and your breath becomes fast and shallow.

How often have you listened to or watched a speaker who really messed up? Probably not very often.

When we have to speak in front of others, we envision terrible things happening. We imagine forgetting every point we want to make, passing out from our nervousness, or doing so poorly we lose our job. However, the reality is that those things almost never happen. We build them up in our minds and end up more nervous than we need to be.

Although these symptoms can be annoying or even debilitating, a certain amount of pressure enhances performance. By changing your mindset, you can use nervous energy to your advantage.

Make an effort to stop thinking about yourself, your nervousness, and your fear. Instead, focus on your objective and the audience. Remember, what you are saying is "about them." You are trying to help, initiate action, or educate the audience in some way. Your message is more important than your fear. Concentrate on the objective and the audience's needs instead of your own.

Visualize yourself being successful, pray, and use deep breathing exercises to slow your heart rate and give your body the oxygen it needs to perform. This is especially important right before you speak. Take deep breaths from your belly, hold each one for several seconds, and let them out slowly.

Crowds are more intimidating than individuals, so think of your speech as a conversation you are having with one person. Although your audience may be 100 people, try focusing on one friendly face at a time, and talk to that person as if he or she is the only one in the room.

Watch Recordings of Your Speeches

You can improve your speaking skills dramatically by watching yourself later and working on areas that didn't go well. Therefore, it is a good idea to record your presentations and speeches whenever possible.

As you watch, notice any verbal stalls, such as "um" or "like." Observe your body language to see if you are swaying, leaning on the lectern, or slouching.

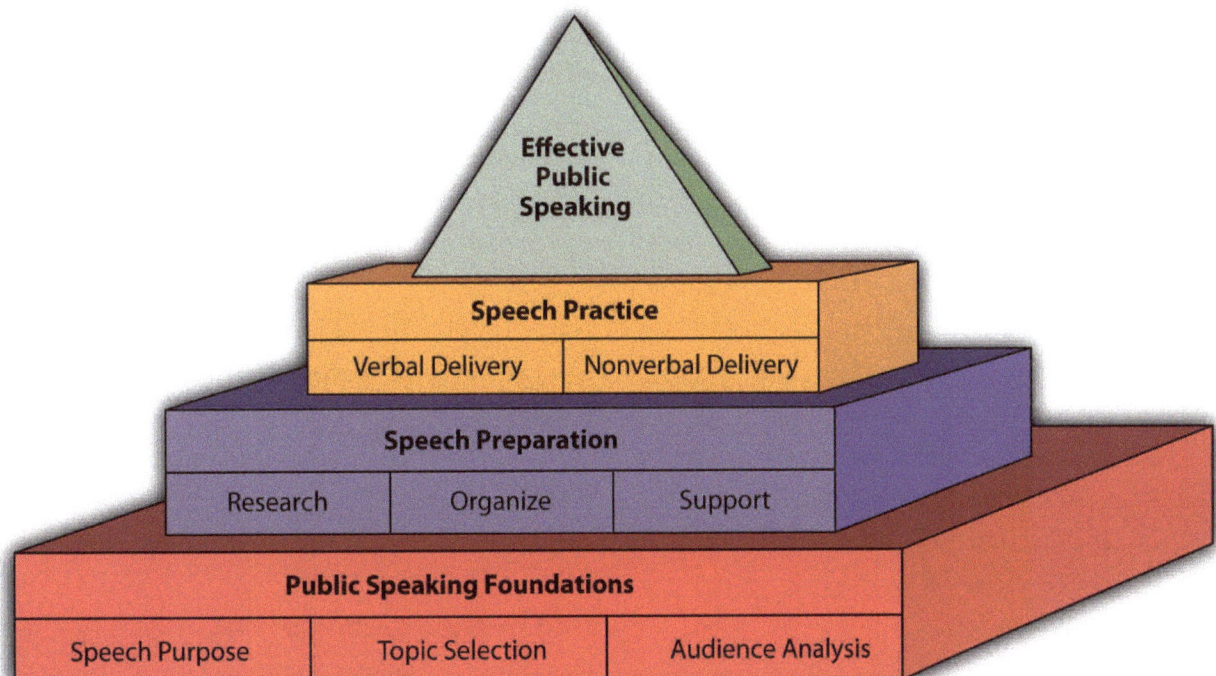

Notice how well you look at the audience, how often you smile, and whether or not you are speaking clearly.

Pay attention to your gestures. Do they appear natural or forced? Make sure people can see them, especially if you're standing behind a lectern.

Finally, look at how you handle interruptions, such as a sneeze or a question you weren't expecting. Does your face show surprise, hesitation, or annoyance? If so, practice managing interruptions like these smoothly, so you're better next time.

Speaking with Passion

The best way to get your audience to care about what you are saying is to show how much *you* care. The best presentations are those that come from the heart.

Emotions bring emphasis to your speech. However, be careful not to get overly emotional. Keep out tears, anger, and overblown elation, or you may be taken less seriously.

Pitch, tone, gestures, and timing will give your speech the semblance of passion; even for those speeches where you don't care about the topic, but you have to talk about it anyway. The most vibrant speakers pause at just the right time, such as before making a valid point.

Don't Rush the Speech

Your presentation is not a race. Take your time as you interact with the audience and slow down if you find yourself hurrying or make a mistake. Many tend to speed up, when they slip up in order to get past the blunder as soon as possible, but that can disrupt your ability to get your message across and often causes more mistakes. Instead, breathe evenly and ease back into your speech with calm confidence.

Speakers often think the audience has a high expectation for a flawless performance, but that is rarely the case. If you make a noticeable mistake, just move on. The audience will forgive you.

Tips to Become a Better Public Speaker

Take a Course—Local colleges and learning centers typically offer courses in public speaking. You will not only learn great techniques, you will be able to practice your skills in front of a roomful of people who are also working to improve their public speaking skills.

Join a Group—Groups like Toastmasters are designed to help professionals hone their speaking skills in front of small groups.

Practice—You don't have to be in a formal setting to practice public speaking. You can practice alone in front of a mirror, with friends, or with family members. Video recording yourself is also helpful.

Prepare—You can overcome public speaking fears through preparation. Do your research and memorize your speech so it will feel natural once you are in front of the audience.

Research Your Audience—Your speaking engagement will be far more effective if your speech is tailored to your specific audience. Spend time learning as much as possible about attendees and prepare your speech accordingly.

Use Tools—PowerPoint has become an essential part of presenting. When used effectively, it can be a great way to keep your speech entertaining.

Know Your Environment—If possible, visit the venue before your speech and familiarize yourself with the environment.

Get Experience—The more experience you get with speaking in front of people, the more you'll improve.

Observe Good Public Speakers—You will get great ideas for your own presentations by watching others who are good at speaking in public. Pay attention to the techniques used by seasoned professionals at conferences and workshops you attend.

Grab the Audience in the Beginning—You will loosen up the audience and develop a rapport, if you can start your speech with something entertaining such as a joke or an interesting story.

Put the Audience to Work—You can take some of the pressure off yourself by gaining audience participation. If appropriate to your speech and the situation, have everyone try an exercise or ask a few people to tell a personal story.

Encourage Questions—As your speech progresses, audience members will inevitably have questions. If conducive to your objective, find ways to get them to feel comfortable voicing those questions, such as rewarding them with prizes.

Focus on the Audience—As you stand up in front of a roomful of people, it can be easy to assume everyone is fully focused on you. This is often not the case. Put your focus on your audience and you will often gain their attention in return.

Dress to Impress—What you wear will make an instant impression. Dress to fit the role you desire to portray.

Avoid Filler Words—One of the biggest complaints audiences have of speakers is the overuse of filler words like "um," "you know," and "so." Eliminate the use of filler words.

Focus on Individuals—A powerful technique recommended by professional speakers is to make eye contact with members of the audience. Try to find a few friendly faces and rotate between them as you speak.

Walk—Pent-up energy can be the death of a powerful speech. It helps to roam the platform or walk around the front of the room while speaking. This helps expend nervous energy and keeps the audience alert.

Breathe—When you are feeling that intense stage fright setting in, use breathing exercises. This activity helps your mind and body to relax.

Use Prompts—Some people write their entire speeches out beforehand, but this can force you to spend the entirety of your presentation reading. Instead write short phrases or bullet points on index cards that will prompt each new idea as you move through your speech.

Team up with a Partner—In your early days as a speaker, don't feel you have to do it all alone. Participate in panels and group presentations where you can share the burden of presenting with a peer or business partner.

Record Yourself—One of the best ways to improve as a public speaker is to record your presentations and watch them later. As difficult and time consuming as this can be, it is a great way to identify where improvement is needed.

Ask for Feedback—As you speak to various groups, allow audience members to anonymously complete a feedback form and use that feedback to improve.

Bring the Right Tools—Be sure you arrive at the site with the tools you will need to conduct your presentation, including wireless clickers, laser pointers, cables, and projector adaptors for your laptop.

Practice Articulation—Work on speaking loudly and clearly. Practice pronunciation of difficult words.

Leave out Slang and Profanity—Slang, profanity, and street talk is not cool. It makes you appear immature, unprofessional, disrespectful of others, and uneducated. Many people are offended by profanity. Keep it real; be considerate of others and be professional

Finish with a Call to Action—Your speech should end with a call to action. What do you want audience members to take away from your presentation? What should they do now?

Final Word on Public Speaking

While public speaking may seem intimidating, the benefits of being able to speak well far outweigh any perceived fears. Speaking well in public can help you get a job or promotion, raise awareness for your cause or organization, motivate others, and/or educate people. The more you speak in front of others, the better you will become at public speaking, while gaining more confidence and speaking with greater ease along the way.

Communication is a powerful tool in all aspects of our lives. It is paramount in business communications, personal communications, and negotiations. The most feared form of communication is public speaking; however, this is also one of the most powerful mechanisms to help you be successful. By utilizing the tips above, you can become a confident public speaker.

CHAPTER 10

Leadership: Characteristics of a Leader

Benefits of Effective Leadership

Why is great leadership so important? Following are the main benefits that come to those who exemplify quality leadership.

1. Great leaders provide **vision**.
2. Great leaders bring **clarity**. This allows people to understand where they are going and makes them feel honored to join the journey.
3. Great leaders **believe. This** gives followers confidence.
4. Great leaders stretch **thinking. This applies to their own thinking as well as** others' and promotes growth in people.
5. Great leaders facilitate improvement of **skills.**
6. Great leaders provide **support. This enables** others to be successful.
7. Great leaders make **hard decisions**. They pay the price so others don't have to.
8. Great leaders take the **bullets**. The buck stops with them.
9. Great leaders expand the **worldview**. They create experiences that help others see things in a new and different light.
10. Great leaders **help others gain more self-respect**.

The world is a better place because of quality leadership.

Leadership Styles

There are as many approaches to leadership as there are leaders. There are also many general styles of leadership, including servant and transactional. Fortunately, businesspeople and psychologists have developed useful frameworks that describe the predominant ways that people lead. When you understand these frameworks, you can develop your own approach to leadership and become a more effective leader as a result.

However, leadership styles are not something to be tried on like so many suits to see which fits. Rather, they should be adapted to the particular demands of the situation, the particular requirements of the people involved and the particular challenges facing the team or organization.

Although leadership is sometimes broken down into many different specific styles, they all fall somewhere within the following leadership style categories.

1. **Autocratic leaders** make decisions without consulting their team members, even if their input would be useful. This can be appropriate

when you need to make decisions quickly, when there's no need for team input, and when team agreement isn't necessary for a successful outcome. However, this style can be demoralizing, and it can lead to high levels of absenteeism and staff turnover.

2. **Participative leaders** make the final decisions, but they include team members in the decision-making process. They encourage creativity, and people are often highly engaged in projects and decisions. As a result, team members tend to have high job satisfaction and high productivity. This is not always an effective style to use, though, when you need to make a quick decision.

3. **Laissez-faire** leaders give their team members a lot of freedom in how they do their work, and how they set their deadlines. They provide support with resources and advice, if needed, but otherwise they don't get involved. This autonomy can lead to high job satisfaction, but it can be damaging if team members don't manage their time well, or if they don't have the knowledge, skills, or self motivation to do their work effectively. (Laissez-faire leadership can also occur when managers don't have control over their work and/ or their people.).

4. **Transactional** leaders provide rewards or punishments to team members based on performance results. Leaders and team members set predetermined goals together, and employees agree to follow the direction and leadership of the manager to accomplish those goals. The leader possesses the power to review results and train or correct employees, when team members fail to meet goals. Employees receive rewards, such as bonuses, when they accomplish goals.

Pros: This method works in most of the cases, where it's applied, provided the employees are motivated by rewards. It is a proven compliance strategy, which works best if the top most leader

in the hierarchy is capable of making most of the important decisions and has a strong personality.

Cons: A powerful and assertive leader will find the transactional model conducive to his way of running things. However, though he will create great followers, he will stunt their growth as leaders. When people get used to doing just what they are told and only as much they are told, they stop thinking 'out of the box'. Original thinking is not really promoted in this system, because of which, an obedient workforce will be created with a lack of imagination. New leaders will be tough to find from the lower strata of power hierarchy. This leadership model will create a stressful work environment. Productivity will be maintained but innovations and breakthroughs will be tough to find. For businesses focused on channelizing the creative potential of their employees, this isn't a recommended model for adoption. It stifles the creative instinct, creating compliant 'workers', but not innovators. Overbearing leaders who exemplify this model, rarely leave behind any proteges or followers, who have it in them, what it takes to inspire and lead people.

5. **Transformational** leadership style depends on high levels of communication from management to meet goals. Leaders motivate employees while also enhancing productivity and efficiency through communication and high visibility. This style of leadership requires the involvement of management to meet goals. Leaders focus on the big picture within an organization and delegate smaller tasks to the team to accomplish goals.

Pros: This effective leadership model will create an enthusiastic work atmosphere and it will drive the organization with innovations. The fact that people are working through self motivation, will certainly guarantee higher output and efficiency. It will naturally develop future leaders from the lot of followers. People will work for the leader, even if the monetary and other benefits offered are lesser, as they will be inspired by his vision. Legacy of transformational leaders creates a line of mentor-proteges, which lead innovative waves in businesses.

Cons: This theory is totally based on the ability of the leader, to inspire the work force to put their best in. Some people places in leaders hip roles do not have the force of character to achieve this leadership style.

Of the leadership styles listed, no one is better than the others; all have a specific time and situation in which they will be the most effective form of leadership. Great leaders are those who select their leadership styles like chefs do knives—based on what is needed at the time, and what is the best tool for the job.

Traits of a Great Leader

What makes a great leader? What do you admire about the people who have led you? What qualities should more leaders possess? What are the most valuable qualities a leader should have? What are the most basic qualities a leader should utilitze?

Leadership and management are terms that are often used interchangeably in the business world to depict someone who manages a team of people. In reality leadership and management have very different meanings. To be a *great manager* you must understand what it takes to also be a *great leader*.

One of the key characteristics of a manager is very basic in the sense that they have been given their authority by the nature of their role. They ensure work gets done, focus on day-to-day tasks, and manage the activities of others. Managers focus on tactical activities and often have a directive and controlling approach. Being tactical is not altogether a negative approach as this is a skill set that is greatly needed in business, especially in the fast-paced environments in which most of us work and live. Being able to organize people to accomplish tasks can be a great asset.

In many organizations, managers are often times the previous high performers at the employee level. Does this mean they are ready for the challenge of people management? In many cases, the answer is no. To acquire the solid characteristics of a manager, these previous high performers must be trained.

When you hear the term leader, a number of images may pop into your head. One phrase that may come to mind is "he or she is a born leader." A common misconception is that leadership is a skill with which people are born. While this can be true, more often than not leadership is a competency is achieved through experience, self-development, and practice.

While managers receives their authority based on their role, a leader's authority is innate in their approach. A commonly coined phrase tells us that leadership is doing the right thing and management is doing things right. This illustrates how the two skill sets need to work together. In order to be fully rounded, you must have the ability to manage the day-to-day tasks and deliver results, while seeing the opportunity for change and the big picture.

Furthermore, studies from multiple business organizations (Fortune, Entrepreneur, Brian Tracy, and Dale Carnegie) show the differences between a good leader and a great one can be boiled down to a specific set of traits. The foundation for being a great leader is building a set of traits that inspire people to follow you and achieve outstanding results.

Fortunately, all these traits are skills you can build over time.

Building and demonstrating these traits does not guarantee greatness. However, the absence of any one of these traits will definitely hold you back from being great. In no particular order, following are some key leadership qualities.

Attributes of Great Leaders

Many leaders are competent, but few qualify as remarkable. If you want to join the ranks of the best of the best, make sure you embody all these qualities all the time. It isn't easy, but the rewards can be truly phenomenal. Following are what leadership professionals consider to be attributes that constitute a great leader:

1. **Optimism** The very best leaders are a source of positive energy. They communicate easily and with enthusiasm. They are intrinsically helpful and genuinely concerned for other people's welfare. They embrace the challenge of finding a solution. They seem to always know what to say to inspire and reassure others. They avoid personal criticism and a pessimistic attitude. They look for ways to gain consensus, and they have a knack for getting people to work together efficiently and effectively as a team.

2. **Focus.** They know where they are going and are on a mission to lead people in order to accomplish the goal. Great leaders have a strong focus and stay the course.

3. **Plan.** Extraordinary leaders plan, and their plans are supremely organized. They think through multiple scenarios and the possible impacts of their decisions, while considering viable alternatives and making plans and strategies--all targeted toward success. Once prepared, they establish strategies, processes, and routines so high performance is tangible, easily defined, and monitored. They communicate their plans to key players and have contingency plans in the event last-minute changes require a new direction (which they often do).

4. **Integrity.** Great leaders have actions, values, methods, measures, principles, expectations and outcomes committed to doing the right thing for the right reason, regardless of the circumstances. People who live with integrity are incorruptible and incapable of breaking the trust of those who confide in them.

5. **Compassion.** Talented people want to work for leaders and organizations who truly care about them and the communities in which they operate. Compassionate leaders are fair situations dealing with crises, encourage followers to better actions and are effective and efficient communicators.

6. **Shared vision and actions.** Great leaders enable people to make positive gains by helping them s understand what is needed and inspiring them to be part of the solution.

7. **Engagement.** Great business leaders are able to engage all team members. They do this by offering them challenges, acquiring their input and recognizing them for their contributions.

8. **Empowering.** Great leaders make their associates feel emboldened and powerful, not diminished and powerless.

9. **Empathy** Extraordinary leaders praise in public and address problems in private, with a genuine concern and desire to help the employeegrow and improve. The best leaders guide employees through challenges, always on the lookout for solutions to foster the long-term success of the organization. Rather than making things personal, when they encounter problems, or assigning blame to individuals, effective leaders look for constructive methods to address the problem and focus on positively moving forward.

10. **Reward and celebration.** A reward may be anything—pay, word of praise, promotion, incentives, bonus, additional leave, etc.—that is given to

employees on the basis of their good personal qualities for their particular jobs and performance. A celebration is the action of marking one's pleasure at an important event or occasion by engaging in an enjoyable, typically social, activity. Rewarding employees or throwing celebrations for the team, develops positive feelings and enhanced morale among team members. Rewards and celebrations give employees gratification and brings success to a business while contributing greatly to the leader and organization as a whole.

11. **Humility.** True leaders have confidence but realize the point at which it becomes hubris. According to research from the University of Washington Foster School of Business, humble people are more likely to be high performers in individual and team settings than their not-so-humble counterparts, and they also tend to make the most effective leaders. Humility in business can be described as a three-part personality trait consisting of an accurate view of the self, modeling teachability, and showcasing followers' strengths. With humility comes more meaningful relationships, better opportunities, and, of course, an increased chance of staying relevant and competitive.

12. **Leverage team strengths.** Effective leaders understand each person's strengths and placeseach person in positions where he/she can flourish and grow.

13. **Collaborative.** Great leaders solicit input and feedback from those around them, making everyone feel part of the process.

14. **Flexible.** Great leaders have the ability to change their plans to match the reality of the situation. As a result, they maintain productivity during transitions or periods of chaos. Leaders skilled at this competency embrace change, are open to new ideas, and can work with a wide spectrum of people. Great leaders treat uncertainty and ambiguity as the new normal. Being flexible includes large changes but also trickles down to everyday activities that are subject to change.

15. **Communicative.** Great leaders make it a point to share their vision or strategy often with those around them.

16. **Solid values.** Great leaders courageously remain true to their values and are consistent in applying them despite opposition and challenges.

17. **Fearlessness.** Great leaders are not afraid to take calculated risks or make mistakes.

18. **Passion.** There are hundreds, if not thousands, of books written about leadership, and anyone can pick them up, skim through them, or even read them closely; however, all the knowledge in the world doesn't make a good leader: this requires passion Passion is the key element. It is the passionate leaders that take calculated risks, step up to the plate, and help make the biggest leaps forward within teams, companies, and organizations. Passion for the projects, passion for the company and passion for the people involved are critical to successful leadership.

19. **Confidence.** Confidence is the cornerstone of leadership. If you don't believe in yourself, no one else will. Not only are the best leaders confident, but their confidence is contagious. Employees are naturally drawn to them, seek their advice, and feel more confident themselves as a result

of their interaction with these leaders. When challenged, they don't give in too easily, because they know their ideas, opinions, and strategies are well-informed and the result of much hard work. But when proven wrong, they take responsibility and quickly act to improve the situations within their authority

20. **Clarity.** The only way to become a great leader is bybeing clear about who you are and what is most important to you. Aspiring leaders fail when they try to become all things to all people, or try to do too much outside their area of excellence. Clarity helps you say "yes" to the right things— and "no" to others.

21. **Care.** The strongest, most effective leaders care— not just about the business, but also about the people in it and the people impacted by it. They show they care through their words and actions. They make sacrifices and work hard.

22. **Honesty.** Great leaders treat people the way they want to be treated. They are extremely ethical and believe that honesty, effort, and reliability form the foundation of success. They embody these values so overtly that their integrity is not questioned.

23. **Decisiveness** Great leaders must make tough decisions. They understand that in certain situations, difficult and timely decisions must be made in the best interests of the entire organization: decisions that require a firmness, authority, and finality that will not please everyone. Extraordinary leaders don't hesitate in such situations. They also know when not to act unilaterally but to instead foster collaborative decision making.

24. **Accountability** Extraordinary leaders take responsibility for everyone's performance, including their own. They follow up on all outstanding issues, check in on employees, and monitor the effectiveness of company policies and procedures. When things are going well, they praise. When problems arise, they identify them quickly, seek solutions, and get things back on track.

25. **Supportive.** Great leaders align rewards and recognition systems to best match their team's profile and deliver results. Great leaders foster a positive environment allowing their team to flourish.

26. **Respect.** Great leaders don't play favorites. They treat all people—regardless of class, seniority, rank in the organizational chart, etc.—the same.

27. **Excellent persuasion abilities.** Great leaders persuade people to work together, to achieve more than they ever thought they could, to reach for apparently impossible goals, to put personal interests aside (at least temporarily) in order to accomplish some larger group purpose. They change minds to push followers into making new decisions and taking action.

28. **Inspiration.** Great leaders communicate clearly, concisely, and often, which motivates others to take action, change course, or do more. They challenge their people by setting high, but attainable, standards and expectations, and they give them the support, tools, training, and latitude necessary to pursue those goals.

CHAPTER 11

Budgeting: Making and Sticking to a Budget

Benefits of Budgeting

Budgeting is important to your present and future well-being. Knowing where your money is going; knowing you will be able to pay your bills, keep the roof over your head, and put food on the table; and having a plan to take care of unexpected expenses will reduce your stress levels and improve your overall health by allowing you to sleep at night. Having a budget will also help you save money, so you can do things you want to do, buy things you want to buy and be comfortable when you retire.

Along those same lines budgeting in business is critical. Budgeting will assist you in staying focused on your business strategy and making your business a success. A budget will create a financial roadmap for your business, allowing you to see how you can get where you want to be from where you are by helping plan for future growth.

What is a Budget?

Simply put a budget is a spending plan and an itemized summary of income and expenses spanning a given period of time. It allows you to allocate your income or business revenue to important items, create an emergency fund for unexpected expenses, along with a savings fund to address future wants or needs.

Why Should Everyone Have a Budget?

No matter how much or how little money you have, you need a budget to prioritize and organize your spending and your goals—both present and future. Creating a budget will help you determine whether or not you have enough money to do the things you want to do. If you don't have enough to do everything you would like to do a budget allows you the ability to prioritize your spending to ensure you are able to do the things most important to you. The following is a specific list of reasons why everyone should have a budget.

- Reach mid and long-term goals—having a budget can help you achieve mid and long term goals such as a new car, new home, expansion of your business and retirement.
- Ensure you don't spend money you don't have—if you have a budget and stick to it, you will always know how much money you earn and how much you can afford to spend each month. This will help keep you from going into debt needlessly and keep you from paying late fees and overdraft charges.
- Help prepare for emergencies and unexpected costs—every good budget should include the development of an emergency fund. This fund

will be there for you if you get laid off from your job, suffer from a debilitating illness or injury that keeps you from working, or go through a divorce or any other major, unexpected event.

- Better sleep habits—If you have a budget, you know exactly how bills will be paid and what you can do each month. This allows you to sleep at night without financial worry.

- Puts you in control—with a budget you control where your money goes as opposed to your money controlling you.

- Better communication—if you have a significant other, business partner or anyone you share finances with, creating a budget together will avoid conflicts and resolve differences on how money is spent.

- Alert you to potential problems—a budget will help you see, if spending in an area is becoming out of control or will start cutting into funds needed in other areas.

- Helps determine if and how much debt you are able to acquire—some debt may be necessary for larger items such as vehicles and homes or to expand your business. Having a budget will help you know how much debt you can comfortably handle.

- Saves Money—having a budget will help eliminate late fees and overdraft fees. These fees add up to significant amounts and eliminating them will save you money.

- Makes forecasting revenues of your business possible—by using previous month's income against expenditures, you are able to predict your profits in the coming months.

- Financial condition readily available—with a budget you have an overview of your assets and liabilities prepared for the government and/or institutions who may be considering offering you a line of credit or loan for your business.

- Forces focus and discipline on your business plan—having itemized and detailed expenditures to measure against income keeps you focused on your business goals.

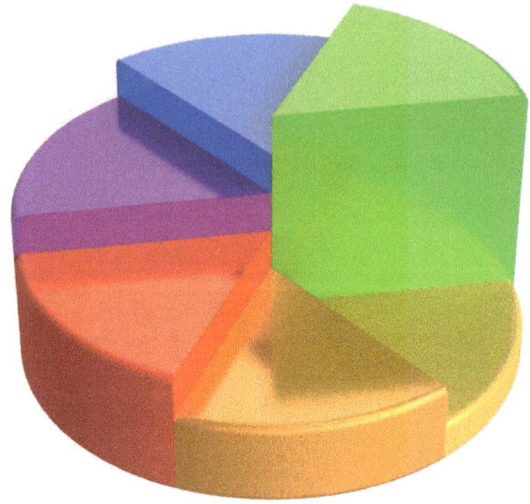

How to Develop a Budget

Everyone needs a budget, but what is the best way to create one? You will need a spreadsheet or other budget tool (electronic, mobile phone app, or pen and paper) to record and track your information (see Budget Tools below), then follow these steps.

1. Gather necessary items—locate and organize your bank statements, paystubs, receipts from purchases, credit card statements, loan documents (personal, vehicle, etc.), utility bills, rent or mortgage information. It is best to gather these for the last three months, if possible.

2. Determine Income—what is your monthly income? Do you receive alimony/child support? Do you have rental income? Receive interest on any dividend, stocks, etc?

3. Determine Mandatory Expenses—these include your rent/mortgage, utility bills, and insurance. Most of these will be fixed in that you know what the amount is each month; however, others may fluctuate month to month. Take the last three months' bills of those that fluctuate, add them together and divide by three to obtain the average monthly amount for that bill.

4. Determine Discretionary Expenses—these include food, clothing, fuel, etc. Food is not optional but the type of food your purchase is. Clothing, also, is not optional, but how much clothing you purchase is at your discretion. Fuel may or may not be optional, depending on your situation; however, the amount of fuel you purchase is discretionary.

5. Determine Optional Expenses—these include items for entertainment and basically unnecessary items, such as tickets to sporting and social events, electronics, etc. Add your receipts from these expenditures for the last three months and divide by three for an average monthly amount you spend in this category.

6. Determine your present situation—When you add your expenses together and subtract it from your income, do you have a surplus, or do you spend more than you receive? If you have a surplus, move on to setting your savings goals (7), debt payoff goals (8) and developing your emergency fund (9).

    If your current spending exceeds your current income, review your optional expenses and see if you can make changes in the amount you are spending there. Then review your discretionary expenses. Can you cut down on dining out? Can you use public transportation instead of driving everywhere? After making these changes, if your expenses still exceed your income, look at the section titled Tips to Save Money.

7. Set Savings Goals—ideally you want to set aside 10% of your income for savings. If you are unable to set aside this much in the beginning, set aside what you can and work up to 10%. Savings should be a line in your budget that you pay each month. If you can have it automatically deposited in this account, you will be less likely to miss it.

8. Set Debt Payoff Goals—What do you want your debt to look like in five years? Ten years? If you have multiple credit card accounts, it is best to pay extra on the one with the highest interest rate while continuing to pay the minimum on all others and move down the list. The object is to get rid of the cards with the highest interest rates; this will save you money.

9. Emergency fund—Ideally your emergency fund should reach an amount equal to three to six months of your living expenses. You will begin this fund by putting aside a little each month based on your situation. You should increase the monthly amount you add to this fund, as you become accustomed to your budget and more money becomes available.

10. Research—If you're beginning a new business, research typical costs and sales trends in businesses similar to yours in your area before creating your budget.

11. Forecast Column—a business budget should include a column that predicts what expenditures and income will be in the coming year.

12. Compare forecasts with actual numbers—When you revisit your budget periodically compare what you predicted numbers would be with what they actually were.

Budgeting Tools

There are many tools available to aid in creating and maintaining a budget. An Excel spreadsheet is a basic way to track your finances; however, some people do not have the time, knowledge, or desire to create the formulas and set up their budget in Excel. Dave Ramsey is famous for his envelope method of budgeting. His method calls for a spending notebook, where you enter your budget categories along with the amounts budgeted to spend in each. Mr. Ramsey then suggests labeling envelopes to correspond with expenditure headings and putting cash (the budgeted amount) in each envelope. This keeps you from spending more than budgeted and allows you to physically see where the cash goes.

If you prefer a more automated, electronic system for budgeting, there are many choices out there. Just find the one with which you are most comfortable to increase the likelihood of sticking to it. Below is a list of a few of the tools available:

1. Mint—www.mint.com states their product allows you to "Effortlessly create budgets that are easy to stick to. We even make a few for you." This application has default categories you can use for your budget, and you can add your own. After you enter your banking information, Mint will automatically track your money, and it also has built-in tools for goal setting. This one is also connected to Turbo Tax, so if you do your own taxes, the information is already there, when you're ready to begin.

2. Level Money—www.levelmoney.com states it "essentially replaces your bank balance with a Spendable number for the month." The "Spendable" number is what you have left after Level Money automatically subtracts your essential bills. Level Money connects to your bank accounts. After you set up spending and savings goals, the app will tell you what you can spend each day by glancing at your phone.

3. GnuCash—http://www.gnucash.org states it is "designed to be easy to use...allows you to track bank accounts, stocks, income and expenses. As quick and intuitive to use as a checkbook register, it is based on professional accounting principles to ensure balanced books and accurate reports." This one includes small business accounting.

4. Pear Budget—https://pearbudget.com states it is "Perfect if you're new to budgeting, or you feel overwhelmed by your finances." This app has some common expense categories to choose from, and you can always change them later if needed. This one has a 30-day free trial but requires a monthly fee after the trial.

5. Centage—http://www.centage.com states its "driver-based budgeting and forecasting ... replaces spreadsheets" for small businesses. This app offers a financial dashboard and reporting.

6. QuickBooks Online—https://quickbooks.intuit.com/online/ states some of its most popular features include tracking expenses, automatically downloading and categorizing bank and credit card transactions, and sending invoices. This tool has a free 30-day trial then requires a monthly fee.

7. FreshBooks—www.freshbooks.com states it is simple to use and will automate tasks such as organizing expenses and tracking time. It also is in the cloud allowing secure access across all your devices. FreshBooks has a free 30-day trial then requires a monthly fee.

Again, the most important thing to remember when choosing a budget tool is to determine what you will use and stick to. For your business you may want to consider an accounting tool that includes invoicing and client/vendor contact information along with budget creation.

Sticking to a Budget

The best advice on how to stick to a budget is to be realistic when developing it. If you dine out every day before developing your budget, do not develop a

budget where you never eat out. You also must have some form of entertainment or hobby in your budget, or you won't stick to it very long. A few other tips on sticking to a budget are listed below.

- Constantly remind yourself of the benefits of your budget. If you're saving for a new car, post a picture of the car in a place where you will see it every day. If one of your goals is to raise your credit score, post a copy of your present credit score. Whatever is motivating you, place a reminder of it where you will see it often.

- Change your mindset. When you consider buying something new, first think about the item you are replacing. Can it be fixed, cleaned, etc?

- Leave your credit card at home. When you go out, only take the amount of cash with you that your budget allows for that expenditure.

- Get rid of high-interest credit cards. If you can get rid of these, you will have more money in your budget, making it easier to stick to it.

- Revise when needed. You should review your budget at least once a month. If some of your goals were unrealistic, adjust them. Don't continue trying to adhere to an unrealistic budget. You will also want to adjust your budget, after you have paid off bills/debt and have more money with which to work or when situations in your life change, i.e., get a new job, get a roommate, get a pet, etc.

- Keep expense categories simple in a business budget. For instance you should have a supply category under expenses as opposed to categories for paper clips, pens, envelopes, etc. individually.

- Learn from your reviews. When you compare your forecasted and actual numbers in your business budget, try determine why the numbers don't match. Are there seasonal reasons? Revise your budget to anticipate these.

Tips to Save Money

There are always ways to cut back on expenses. A few people don't always think of are listed below.

- Limit dining out—cut back on the number of times you eat out each week. If you typically go out for lunch which averages $10 multiplied by five work days in a week and four weeks in a month, that is $200 spent on lunch every month. Consider taking your lunch more. An average evening meal costs $20. If you typically eat out four nights a week, you are spending $320/month on evening meals. Cook more at home.

- Make your own coffee—a plain, regular cup of coffee may cost you $2.00. If you stop for this cup of coffee every morning on your way to work, you are spending $40/month and $480/year on coffee. Try making your own coffee at least part of the time.

- Grocery list—make a list before you go shopping and focus on that list. Resist the temptation to add to your list while shopping.

- Shop when on a time schedule—rush yourself. Try to buy groceries, when you know you have to be somewhere else at a certain time. Don't give yourself time to add to your shopping list or wander around the store picking up excess items.

- Check expiration dates—This will keep you from wasting food, which is wasting money.

- Buy in bulk—nonperishable items. Buying in larger quantities is often cheaper in the long run but confirm this by comparing prices and calculating the cost of the individual items.

- Buy generic—many items have generic forms with little to no difference in taste or effectiveness.

- Avoid late fees—pay bills on time.

- Avoid ATM fees—many banks charge a fee when using ATMs other than theirs, and the owner of the machine charges an additional fee. Therefore, you are being charged two fees to have access to your own money. Be sure you use one of your bank's ATMs or withdraw cash directly from your bank.

- Avoid impulse buying—especially on large items. When you find something you want to purchase, wait two days before making the actual purchase. Many times you will realize it is not something you really need.

- Avoid vending machines—vending machines typically have a large markup on items. Bring snacks from home to keep in your desk.

- Cut your phone bill—if you still have a landline, do you use it? There is typically no reason to have a landline and a cell phone. Review your cell phone bill. Is there a better plan for your situation? Are you exceeding your data limit which causes extra charges?

- Cut your satellite/cable bill—is there a better plan for you? Review your bill and the channels for which you pay. You may be able to decrease the number of channels you receive and not even miss them. If you typically only watch movies, it may be a good idea to get rid of your cable/satellite and subscribe to Netflix or something similar for less money than your current cable/satellite.

- Get rid of your fax machine—if you still use a fax machine in your business, determine if you really need it. In most cases, you can scan and email anything you previously faxed.

- Be sure you are getting the best internet deal— shop around for the best internet prices. Of course you also have to consider quality of service here.

It won't do you any good to save money on your internet, if it slows down your work.

- Look at office supply costs—visit with different suppliers to determine who will give you the best rate. Some may cut costs for small businesses. Is delivery extra? It may be worth it to pick up supplies, if you buy locally.

Budgeting Success

Many people think they don't have enough money to budget, budgeting is too much of a headache, or budgeting will cut out all their fun. The truth is budgeting will help you have more money by paying off debts and avoiding unnecessary fees. While developing a budget does take a little time, it is not difficult and sticking to your budget gets easier as you go along. It also gets to be more fun, because you know where your money is going and begin having more of it to do the things you want. Even when beginning your budget you should always allow for entertainment expenses. Budgeting is a win-win as you gain more control of your life, have the means to do the things important to you, become more financially stable, and experience less stress.

An effective budget will also allow your business to grow. When following a budget, you save money on expenditures which can be placed in a reserve fund for future expansion. Budgeting for future growth like this not only makes the growth possible but ensures the money is available at any time an opportunity may arise for growth or expansion.

CHAPTER 12

Persuasion

Definitions

Persuasion is the action or fact of persuading someone or of being persuaded by someone to do or believe something.

In other words, persuasion is the process of changing or reforming attitudes, beliefs, opinions, or behaviors toward a predetermined outcome through voluntary compliance. Persuasion is not the same as negotiation, a term that suggests some degree of backing down or meeting in the middle. Rather than compromising, as in negotiation, effective persuasion will actually convince the opposing party to embrace your view.

Influence is who you are and how you, as a person, will impact the message. This includes whether or not you are viewed as a trustworthy and credible person.

Power is the possession of control or command over others. Authority is power. Power increases your ability to persuade and influence. This power can be seen in people who possess knowledge, have authority, or use coercion during a persuasion process.

Motivation is the ability to incite others to act in accordance with the suggestions and ideals you have posed. Motivation is a "call to action" or the urge to do something.

The Importance of Persuasion

Whether it is taking your business to the next level, cajoling your boss into giving you a raise, winning someone round to your point of view, or persuading your spouse to put out the trash, getting people to do what you want can be very handy.

Persuasion, a basic form of social interaction, is a key element of all human interaction from politics to marketing to everyday dealings with friends, family and colleagues. The power of persuasion is of extraordinary and critical importance in today's world. Nearly every human encounter includes an attempt to gain influence or persuade others to our way of thinking. Regardless of age, profession, religion, social standing, or philosophical beliefs, people are always trying to persuade each other. We all want to be able to persuade and influence others to listen to, trust, and follow us. A recent study by economists found that 26 percent of the gross domestic product was directly attributable to the use of persuasion skills in the marketplace.

Not only does persuasion drive our economy, it is also the key ingredient to a company's success. Rarely do large corporations downsize their sales forces. Sales professionals are assets to the company, not liabilities. Top notch persuaders will always find employment, even in the slowest of economies.

Whether used for good or for bad, the ability to persuade is power. Persuasive people keep kids off drugs, prevent wars, and improve the lives of others. Unfortunately, persuasive people also get kids on drugs, stir up wars, and destroy lives. Obviously, the focus here is to understand persuasion for the improvement and betterment of ourselves, our friends and families, and our communities while also being able to identify the persuasive influences upon our lives.

The notion that being a refined persuader means being forceful, manipulative, or pushy is completely wrong. These tactics may get short-term results, but maximum influence is about getting long-term results. Lasting influence isn't derived from calculated maneuvers, deliberate tactics, or intimidation. On the contrary, proper implementation of proven persuasion strategies will allow you to influence with the utmost integrity leading people to naturally and automatically trust you, have confidence in you, and even want to be persuaded by you.

It is a common misconception that only individuals involved in sales, marketing, or leadership positions need to learn persuasion skills. Everyone needs persuasion skills, no matter their occupation, and everyone uses the techniques and tactics of persuasion each day. All of us try to get others to do what we want them to do or to change their view. Therefore, mastering communication and understanding human nature are essential life lessons, if we want to effectively persuade and influence people.

We can't get anywhere in life without the ability to work with other human beings. It is through our dealings with others that we achieve success, because no one is self-sufficient. Everything of any value in life is achieved through the support and help of the people around us. As a society, we are interconnected, and the ability to make positive connections and be persuasive is vital to our success.

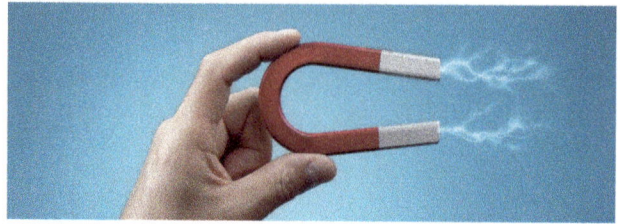

Being Alert to Persuasive Influences

Advertisers spend billions of dollars researching and analyzing our psycho-graphics and demographics to figure out how to persuade us. Research has shown that the average person is exposed to 300 to 400 persuasive media messages a day from the mass media alone. We are bombarded with thousands of persuasive messages through a myriad of sources, including newspapers, magazines, billboards, signs, packaging, the Internet, direct mail, radio, TV, mail order, catalogs, coworkers, management, sales professionals, parents, and children. The problem is: Thousands, even millions, are persuaded against their better judgment every day, simply because they are unequipped to identify the persuasive concept, accurately interpret the message, and effectively respond to the advertising barrage we perpetually face. In this case, what you don't know will hurt you. Persuasive influences flood our daily existence and are inescapable. Without question it is in our best interest to master the subject of persuasion, know how it works, and learn how to implement its proven techniques to empower us today.

Tools for Success

There is no question that we want and need things from other people. We also have to admit we want people to follow, trust, and accept us. And if we are truly honest we have to confess that we want to influence others to our way of thinking. Bottom line: we want to get what we want — when we want it. Possessing the right tools and knowing how to use them (for good) is the secret to success.

Most people use the same limited persuasion tools over and over achieving only temporary, limited, or even undesired results. However, you can do only so many things with a hammer. For the best results and maximum success we need to learn all the techniques for persuasion and influence. To be most effective, persuasion must be customized to the group or individual and be appropriate for the situation.

Psychologists have long been fascinated by persuasion—studying why some people are more persuasive than others and why some strategies work where others fail. This chapter will address the science of persuasion and show you proven methods that work.

Persuasion and Rhetoric

In the ancient Greek world one's ability to persuade meant great social prestige. It was Aristotle who first introduced persuasion as a skill that could be learned. At that time, rhetorical training became the craze for the citizens of Athens, especially the politically elite. The first book ever written on persuasion was Aristotle's The Art of Rhetoric. The book's basic principles established a foundation for persuasion that still holds true today.

Aristotle taught that rhetoric was an art form that could be approached systematically by a formula for all persuasive attempts. Aristotle's most famous contribution to persuasion was his three means of persuasion: ethos, pathos, and logos. He argued that the most effective persuasive attempts contain all three concepts, setting an unshakable foundation for success. A brief review of Aristotle's three basic means of persuasion are provided below.

Ethos

Ethos refers to the personal character of the speaker. Aristotle believed audiences could be persuaded if they perceived a speaker as credible. In his own estimation: "We believe good men more fully and readily than others." Aristotle also stated that "ethos is not a thing or a quality but an interpretation that is the product of the speaker-audience interaction." Ethos includes such things as body type, height, movement, dress, grooming, reputation, vocal quality, word choice, eye contact, sincerity, trust, expertise, charisma, etc. It is the audience's perception of the credibility of the speaker.

Aristotle taught that ethos was the most powerful of the three persuasive means. Indeed, scientific research has proven the power of individual ethos. A study by Hovland and Weiss, professors at Yale University, gave students messages that were identical in all respects except for their source. High-credibility sources yielded large opinion changes in the students, while low-credibility sources produced small opinion changes.

Pathos

Pathos is the psychological state of the audience. The psychological or emotional state of the listener can affect persuasion because "our judgment when we are pleased and friendly [is] not the same as when we are pained and hostile." When considering pathos, it is important to know both the individual's actual state of mind and his desired state of mind. When you determine the difference between the two, you can use that knowledge to your advantage. The belief is that by helping them see how they can get from their current state to their desired state, you can persuade people to do just about anything.

Logos

Logos is the substance of a message, or the logic presented to provide proof to the listener. Aristotle believed that humans are fundamentally reasonable people who make decisions based on what makes sense. This manner of reasoning is what enables the audience to find the message persuasive and convincing.

Aristotle's three concepts are central to understanding modern-day persuasion. The principles and laws are founded upon proven principles presented by Aristotle and the ancient Greeks. Although the times and the means of persuasion have changed over the years these principles remain the foundation for persuasion.

Modern-day persuaders run into three major factors that make persuasion a greater challenge than it was in the past. First, people have the ability to rapidly obtain needed information via today's technology. Through the use of the Internet and telephone, information is instantly available. We can now find the cost of a car, before we even enter the dealership. The second roadblock to persuasion is that today's consumers are increasingly doubtful and skeptical. The number of persuasive arguments we see and hear every day is growing at an alarming rate, and it takes more and more effort to sort out the valid offers from the scams. People are having to tune out many persuasive messages or be overwhelmed; therefore, they are on guard against messages directed at them. The third barrier to persuasion is choice. Now, via the Internet, the consumer has access to the world market. In the past, if you had the only bookstore in town, that is where people had to shop. Now, one bookstore owner has to compete with hundreds of competitors around the globe for the same business.

Persuasion, Communication, and Knowledge Leads to Confidence

The greatest common denominator of the ultra-prosperous is they are master communicators. There is a direct correlation between your ability to persuade others and the level of your income. Impeccable and masterful communication unarguably leads to wealth. The highest paid and most powerful people on the planet are all master communicators. These individuals put themselves at stake in front of large groups, communicating and persuading in such a way that people are inspired to support them. Your financial success in life will be largely determined by your ability to communicate with other people. Everything you want, but don't currently have, you will have to get from others. Your ability to effectively communicate and persuade will be your key to riches.

Persuasion is also your golden ticket to promotion. Communication skills rank number one of all the personal qualities employers seek in college graduates. While most people shy away from overtly persuasive situations, master communicators welcome such opportunities. Master communicators feel in control of challenging situations, because they understand the art of persuasion, and they know how to recognize and use persuasive strategies.

Persuasive Techniques that Work

Getting people to do what you want and, at the same time, enjoy it is not an accident or coincidence. You must use techniques based on the proven methods of persuasion and influence to achieve such results. As you master these techniques, you will positively influence others and be rewarded with greater success.

Professional negotiators, successful sales professionals, and influential business leaders around the world use these techniques. They are the same methods that help thousands of people gain control of their lives and their financial futures. The more you learn and apply these techniques the greater your persuasive influence will be upon others, and it will not be long before you find yourself in a completely different position than you are today.

10 Psychological Theories to Persuade People

1. Amplification Hypothesis

When you express with certainty a particular attitude, that attitude hardens. The opposite is true as well: Expressing uncertainty softens the attitude.

2. Conversion Theory

The minority in a group can have a disproportionate effect on influencing those in the majority. Typically, those in the majority who are most susceptible are the ones who may have joined, because it was easy to do so or they felt there were no alternatives. Consistent, confident minority voices are most effective.

There are four major factors that give the minority its power:

1. *Consistency*: Being consistent in expressing minority group opinion.
2. *Confidence*: Being sure about the correctness of ideas and views presented.
3. *Unbiased*: Appearing to be reasonable and unbiased in presenting ideas.
4. *Resistance*: Resisting the natural social pressure and abuse the majority may bring to bear on minority members.

In addition, to gain the confidence of the 'silent majority,' the minority shows that it is *not* like the leadership of the majority, typically by visibly opposing them (something most of the silent majority would not dare do). They then show empathy and similarity with the target people, steadily subverting them and convincing them to join their alternative group.

Conversion Theory Examples:
- A business executive board is keen to acquire another company, although the decision is mostly being driven by the CEO and CFO. There seems to be consensus on this, but the CTO thinks it is crazy. He asks public and challenging questions about the move whilst talking quietly to other board members until he is confident he can call a motion of no confidence in the move.
- An extremist group holds regular demonstrations against the local government but does this peacefully, engaging people passing by in reasonable and persuasive conversation, getting them to sign a petition and possibly come to the next meeting.

3. Information Manipulation Theory

This theory involves a persuasive person deliberately breaking one of the four conversational maxims. These are the four:

- Quantity: Information is complete and full.
- Quality: Information is truthful and accurate.
- Relation: Information is relevant to the conversation.
- Manner: Information is expressed in an easy-to-understand way and non-verbal actions support the tone of the statement

4. Priming

You can be influenced by stimuli that affect how you perceive short-term thoughts and actions. Following is an example:

A stage magician says 'try' and 'cycle' in separate sentences in priming a person to think later of the word 'tricycle.'

5. Reciprocity Norm

A common social norm, reciprocity involves our obligation to return favors done by others. The norm of reciprocity is the expectation that people will respond favorably to each other by returning benefits for benefits, and responding with either indifference or hostility to harms. The social norm of reciprocity often takes different forms in different areas of social life, or in different societies.

6. Scarcity Principle

The scarcity principle is an economic principle in which a limited supply of a good, coupled with a high demand for that good, results in a mismatch between the desired supply and demand equilibrium.

7. Sleeper Effect

The sleeper effect is a psychological phenomenon that relates to persuasion. It is a delayed increase of the effect of a message accompanied by a discounting cue. Persuasive messages tend to decrease in persuasiveness over time, except messages from low-credibility sources. Messages that start out with low persuasion gain persuasion as our minds slowly disassociate the source from the material (i.e., a presumably sleazy car salesman and his advice on what car is best).

8. Social Influence

We are influenced strongly by others based on how we perceive our relationship to the influencer. For example, social proof on web copy is persuasive if the testimonials and recommendations are from authoritative sources, big brands, or peers.

Social influence occurs when one's emotions, opinions, or behaviors are affected by others. Social influence takes many forms and can be seen in conformity, socialization, peer pressure, obedience, leadership, persuasion, sales and marketing.

9. Yale Attitude Change Approach

The Yale attitude change approach, also known as the Yale attitude change model, is the social psychology study of the conditions under which people are most likely to change their attitudes in response to persuasive messages. This approach, based on multiple years of research by Yale University, found a number of factors in persuasive speech: being a credible, attractive speaker; when it's important to go first or go last; and the ideal demographics to target.

10. Ultimate Terms

Certain words carry more power than others. This theory breaks persuasive words into three categories:

- God terms: those words that carry blessings or demand obedience/sacrifice, e.g., progress, value
- Devil terms: those terms that are despised and evoke disgust, e.g., fascist, pedophile
- Charismatic terms: those terms that are intangible, less observable than either God or Devil terms, e.g., freedom, contribution

Psychological Theories to Practical Application

The above 10 Psychological theories can be considered the building blocks for persuasive techniques used today (referenced by different names) and explained below. With this foundation of psychology in place, application in the real world is more effective.

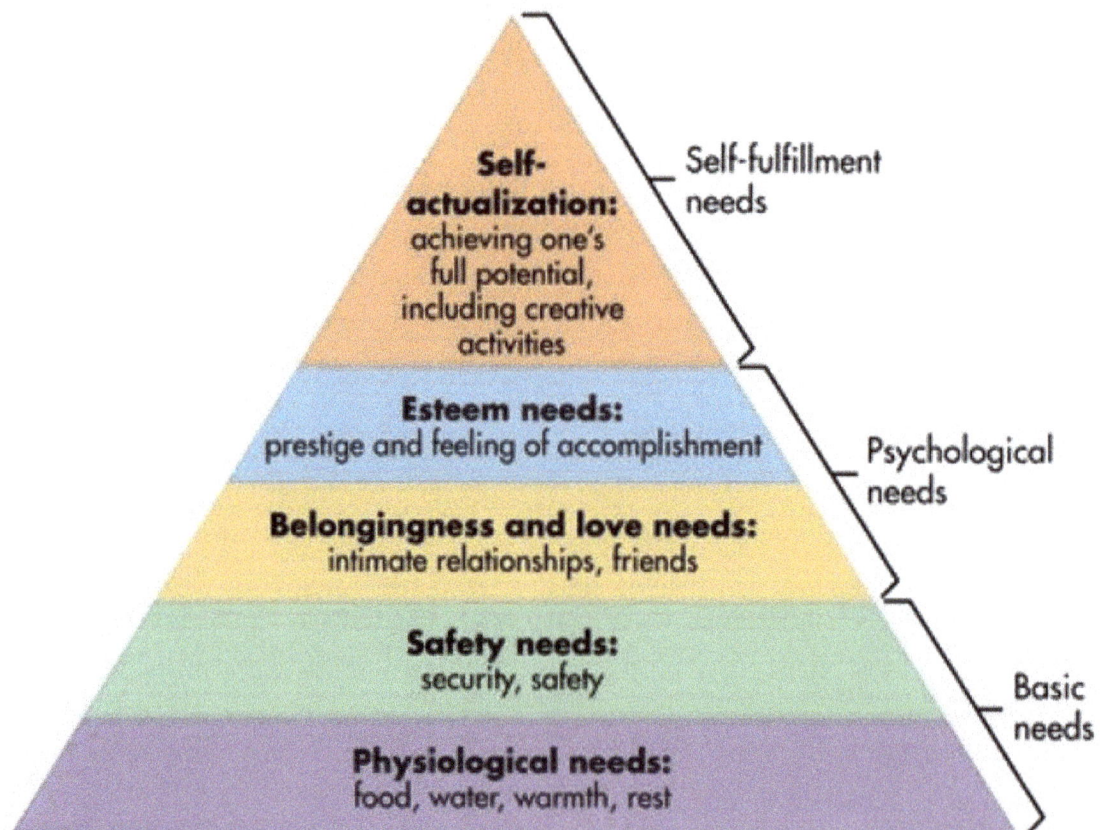

Self-actualization: achieving one's full potential, including creative activities — Self-fulfillment needs

Esteem needs: prestige and feeling of accomplishment

Belongingness and love needs: intimate relationships, friends — Psychological needs

Safety needs: security, safety

Physiological needs: food, water, warmth, rest — Basic needs

Persuasion through Another's Needs and Cravings

The Hierarchy of Needs pyramid, proposed by psychologist Abraham Maslow in the 1940s, shows the advancing scale of how our needs lay out on the path to fulfillment, creativity, and the pursuit of what we love most. The version of the pyramid you see below shows the five different layers of needs.

The three steps in between the physiological needs and the fulfillment needs are where persuasion most directly applies.

- Safety
- Belonging
- Esteem

In Maslow's pyramid, the descriptions for these needs don't exactly have a marketing perspective to them, but with a little creativity it is easy to see how you can tailor your message to fit these needs.

Without these three essential keys it is difficult, if not impossible, for a person to be innovative, emotionally engaged, or move forward to the next level. The more feelings we have of these three keys, the greater the success of the company, the relationship, the family, the team, and the individual.

Zig Ziglar was one of the top sales persons in several organizations before striking out on his own as a motivational speaker and trainer. His motto was "You will get all you want in life if you help enough other people get what they want." In other words, if you want to achieve your goals, help others achieve theirs.

How to Win Friends and Influence People

Dale Carnegie's most famous work, "How to Win Friends & Influence People," was published in 1936, but its insights into human nature and persuasion are as relevant today as they were then. Summarized below are his tips on how to persuade even the most stubborn people.

1. Don't try "winning" an argument.

Even if you manage to tear apart someone else's argument, you don't actually achieve anything. Carnegie cites the old saying, "A man convinced against his will is still of the same opinion." If you are trying to persuade somebody, avoid an argument.

2. Respect other people's opinions.

Pride — both yours and the person's you're trying to convince of something — is the biggest impediment to reaching an agreement. Be diplomatic about presenting your opinion, Carnegie explains, and never say, "You're wrong," no matter how true it may be.

3. Admit when you're wrong as soon as you realize it.

"When we are right, let's try to win people gently and tactfully to our way of thinking, and when we are wrong — and that will be surprisingly often, if we are honest with ourselves — admit mistakes quickly and with enthusiasm," Carnegie writes. It will allow both you and whoever has pointed out your mistake to clear the air and move on.

4. Be friendly, no matter how angry the other person may be.

It's human nature to meet aggression with aggression, but if you take the high road and try to persuade someone while maintaining a smile and showing appreciation for their situation, you'll be surprised what you can achieve.

5. Reach common ground as soon as possible.

"Begin by emphasizing — and keep on emphasizing — the things on which you agree," Carnegie writes. "Keep emphasizing, if possible, that you are both striving for the same end and that your only difference is one of method and not of purpose."

6. Let the other person do most of the talking.

The average person enjoys speaking about him/herself more than any other topic, and if you're engaging someone who has a lot to say, they're not going to listen to you until they've put it all out there. Listen more than you speak.

7. Get the other person to think your conclusion is their own.

No one can be forced to truly believe something. That's why the most persuasive people know the power of suggestions over demands. Plant a seed, water it, and when it blossoms, avoid the urge to take credit for it.

8. Figure out why the other person thinks what they do.

The person you're trying to convince can be objectively wrong about something, but they believe what they do for a reason, and that doesn't necessarily make them a bad person. "Ferret out that reason — and you have the key to his actions, perhaps to his personality," Carnegie writes.

9. Determine how their beliefs work in their favor.

Behind every closely-held opinion is a lifetime that has led to that person's conclusion. It's in your interest to sympathize with how the belief in question fits into the other person's worldview, and how that worldview is a complex machine driving the other person through life.

10. Appeal to nobler motives.

Carnegie says everyone but the most bitter or stubborn among us actually wants to do what they consider to be the right thing. Frame your argument with morality.

11. Be dramatic.

If you have truth on your side, make it as appealing to emotions as you can.

12. When nothing else works, "throw down a challenge."

If you truly can't convince another person to do or believe something, then appeal to their competitive side. Challenge them to either prove why they are correct, or if you're a manager, challenge your employees to do something to prove their worth.

Other Opportunities for Persuasion

So far we have looked at persuasion from a personal standpoint (communication to another person or small group of people). However, applying the principles of persuasive psychology can be used on larger audiences via the following methods:

- Calls-to-Action
- Headlines in Articles
- Tweets and Updates
- Bulk Emails
- Product Descriptions
- Websites

Almost anywhere you have words or visuals, basically anywhere you create or manage content, you have an opportunity for persuasion.

Persuasion through Storytelling

People can often be more easily persuaded by hearing stories than by being recited facts. This means one way to persuade someone to change their minds is by offering an alternative story.

Three common story types proven to be effective persuasion methods are:

1. **The Challenge Plot**: A story of the underdog, rags to riches or sheer willpower triumphing over adversity
2. **The Connection Plot:** A story about people who develop a relationship that bridges a gap whether racial, class, ethnic, religious, demographic or otherwise; think of the film *The Blind Side*
3. **The Creativity Plot:** A story involving someone experiencing a mental breakthrough, solving a long-standing puzzle or attacking a problem in an innovative way

Additional Effective Persuasion Techniques

The following methods have been proven to be effective means of persuasion. However, they should only be implemented ethically and never used to manipulate others.

Be a Mimic

People can become very irritated, if they discover someone is mimicking them. Yet a number of recent studies have concluded that mimicking someone's mannerisms subtly—their head and hand movements, posture and so forth—can be a powerful form of persuasion. For example, William Maddux at the INSEAD business school in Fontainebleau, France, explored the effect of mimicry on 166 students in two role-play experiments, one involving negotiations between job candidates and recruiters, the second between buyers and sellers.

In both cases the outcome of the negotiations was better for the would-be persuaders when they employed subtle mimicry. For example, in the buyer-seller experiment, 67 percent of sellers who mimicked their target secured a sale, as opposed to 12.5 percent of those who did not.

However, be careful as overt mimicry can backfire or be very embarrassing if detected. The crucial factors are to be subtle, utilize a delay and cease mimicking if you think there's even the slightest chance they are on to you.

2. A Different Perspective

If you want to bring people round to your point of view try "framing," a favorite tactic of spin doctors. Framing is about leading people to think about an issue or opinion in a way that is advantageous to you. Framing is a key tactic in election campaigns.

In one study 69 undergraduates were asked to read an article about two fictitious candidates' views and policies. One candidate Rick was conservative, while the other Chris was liberal. Half of the students had to choose between two statements: "I support Rick" or "I oppose Rick"; the other half had to choose between equivalent statements regarding Chris. Participants also rated their preferences for both candidates on a sliding scale from "strongly support" to "strongly oppose."

They then read another article arguing against the merits of their preferred candidate and then their opinions were reassessed. Overall, people whose preference for a candidate had been expressed in terms of opposition to the other candidate were less likely to modify their opinions. A simple change in framing—leading people to think of their evaluations in terms of whom they oppose instead of whom they support—leads to stronger, more resistant opinions.

These findings fit with a broad body of research suggesting negative information frequently has a more powerful influence than positive messages. So, if you want to sway someone when they choose between two options, an effective tactic is to be negative about the option you don't want them to pick.

3. Less is More

In most battles, outnumbering your opponent will hand you victory, and it would seem common sense that the more arguments you use, the more persuasive you'll be. Yet, the evidence suggests otherwise. A number of studies have revealed that the more reasons people are asked to come up with in support of an idea, the less value they ascribe to each. The result: asking people to "think of all the reasons why this is a good idea" is likely to backfire and may serve to harden their views.

A joint study between Ohio State University and the Autonomous University of Madrid in Spain, demonstrated this effect in 2002. Researchers told 59 university students there was a plan to introduce new exams into their courses—an unwelcome prospect. They then asked half the students to produce two reasons why this was a bad idea, and the other half of the students were asked to produce eight reasons. They were also asked to rate their level of opposition. On average, students who supplied just two arguments against the proposal were subsequently more opposed to the exam policy than those who gave eight.

In conclusion, it was stated that the ease with which we can summon thoughts affects how much confidence we place in them, and it is generally easier to think of two reasons for believing something rather than eight. This finding has some clear practical implications. If you want to persuade people by getting them to think positively about your message, idea, product or whatever, ask them to generate just a few positive thoughts—three at most—because that's easy, and they'll feel confident about their positive thoughts. In summary, when you are trying to persuade someone, avoid the temptation to say "Give me one good reason (not to think the way you want them to)…."; as it will likely only strengthen your opposition.

4. Grind them Down

Hunger is a powerful thing, but how many times have you reached for a quick snack, only to regret eating way too much? Just as your defenses for food quality can slip when your stomach is empty, so can your defenses when in engaging in argument or doing battle with sales people with your mental batteries running low. Conversely, if you're trying to be persuasive, strike when your target is running low on mental energy. Reliable data, via numerous studies, shows there is a decreasing level of resistance to persuasive influences as mental, physical or emotional fatigue increases.

5. Man to Man and Woman to Woman

In this fast-paced world, we seldom have time for face-to-face meetings. You are just as likely to conduct your personal and business negotiations by email, or some other electronic medium, as you are in person. How does this impact your powers of persuasion?

This question intrigued Rosanna Guadagno of the University of Alabama and Robert Cialdini of Arizona State University who have been comparing the persuasive powers of online communication and face-to-face meetings.

In a study published in 2002, Guadagno and Cialdini had a group of students discuss the introduction of new exams. The group was split into same-sex couples. Unbeknown to the subjects, each pair included an accomplice of the experimenters whose role was to provide arguments in favor of the idea. Half the discussions took place in an online chatroom; the other half sat face-to-face.

Overall, men rated the proposals similarly whether they participated in the electronic or face-to-face sessions; however, women in face-to-face sessions rated them more highly than women who only took part online. Guadagno and Cialdini suggest this is because groups of women tend to form communal bonds and reach agreement. Electronic communication disrupts the exchange of social cues women use to establish a communal bond and is therefore less conducive to persuasion.

On the other hand, groups of men typically try to establish their competence and independence, which can lead to competitive encounters. When two men who have not met before debate a point, online interaction is about as effective and persuasive as face-to-face. However, if they have met and had a competitive exchange, subsequent face-to-face meetings are less productive, whereas online exchanges fare far better.

While online communication can prevent women from "connecting," it can help men suppress competitive urges that hamper persuasion. Therefore, if you're a woman and want to persuade other women you'd be better off meeting face-to-face, but a man wanting to persuade another man may fare better with online communication as men are less confrontational if contacted by email.

7. Anger

Angering people may seem like an odd way to go about persuading them, but according to Monique Mitchell Turner, a communications professor at the University of Maryland, College Park, it is seriously underrated as a tool of persuasion.

Anger is typically focused on a perceived injustice And is a natural response that often makes people feel empowered.

There has been a long debate, she says, about whether anger can be constructively harnessed. In study groups that employed anger as a tactic—most notably animal rights groups such as People for the Ethical Treatment of Animals, as well as environmental organizations and even political campaigns—Ms.

Turner found that, given the right conditions, anger can be an effective tool.

First, people have to be convinced the issue is relevant to them—it affects them or their children or their community. At that point, says Turner, you need to hammer home what's wrong with the world as it is. Once you have people riled up, you can offer them a way to remedy the situation. When those feelings of anger are accompanied by the feeling that there is a solution to this problem, then the message is more likely to be persuasive.

8. Resistance

Historically, psychologists studying persuasion have concentrated on what makes certain messages more appealing than others. Over the past few years, however, researchers have begun looking more at the presenter than the message. A growing body of evidence suggests that breaking down people's resistance to persuasion can be even more important than the delivery method.

People are naturally suspicious of attempts to persuade them, especially if they think it is a persuasive message.

Numerous studies have shown that merely reminding people they are vulnerable to manipulation makes them generally more difficult to persuade. Furthermore, people who successfully resist persuasion often become even more entrenched in their opinions, and the stronger they perceive the attempt at persuasion to be, the more certain of their opinions they become and less susceptible to persuasion.

It would seem as though a strong authoritative argument would hold greater sway. However, that is not always the case. It seems that if people resist good arguments presented by an expert, they conclude their own arguments must be even stronger. The recommendation is to present positions closer to your target's views, and then move them towards your goal a little at a time.

9. Style over Substance

It was midnight when the knock came at the door. It was "Paul," a "neighbor" who had "just moved in." He spoke non-stopdetailing a problem with a truck that had run out of gas and his need for $20, which he would, of course, return first thing in the morning. Later, Kurt often looked back and wondered just how it was he had been taken in so easily.

"Paul" was a master of his craft: Kurt later learned that four other people on the street had also been taken in by the con.

Maybe we shouldn't be so surprised when things like this happen. Persuasion may have as much to do with how you say something as what you're saying, and the less time one is allowed to think about the content, the more the style of delivery matters. At least, those are the findings of two marketing professors who decided to tease style and substance apart.

John Sparks at the University of Dayton in Ohio and Charles Areni at the University of Sydney, Australia, knew from earlier work with courtroom transcripts that people equated certain kinds of speech with lack of credibility. For instance, hesitant phrases such as "I mean," "you know" and "isn't it?" reduced a speaker's power.

The University of Dayton study, however, looked at the exact relationship between style and content. The researchers asked 118 undergraduates to read a transcript of a testimonial regarding a scanner. In one version, the speaker used hesitations like "I mean" and "ummm"; in the other, he used none. They also gave half the students enough time to read it thoroughly, while the other half got just 20 seconds, to see how limiting a person's understanding of the substance may alter the persuasiveness of the style.

The researchers found that in both versions style was important. When hesitant language was used, people were less easily convinced this was a scanner worth buying—even when it was a better scanner at a lower price. Style was especially important, the researchers found, when time was limited. They found that most often when people can't pay attention to what the speaker is saying, they focus their attention on how the message is being communicated (i.e., how they say it).

In conclusion, if you want to be persuasive, don't stumble, pause or use language that indicates any kind of hesitation, and limit the time your listeners have to think about what you're really saying.

10. Using Numbers to Your Advantage

A *Scientific American* article titled *Why Things Cost $19.95* showed that using precise numbers when pricing items for sale was more persuasive than using round numbers. The author cited a study which spanned five years of real estate sales in Florida and compared list prices with the actual sale prices of homes. The study revealed homeowners who listed their homes more precisely—say $494,500 as opposed to $500,000—consistently received an amount closer to their asking price. When faced with a precise asking price, buyers were less likely to negotiate . Therefore, when you price your services or products don't use round numbers.

12. The Power of Pink

Color is a powerful force that can affect how we think, feel and behave without our conscious awareness. Pink—or, more precisely, Baker-Miller Pink, also commonly known as Drunk Tank Pink —is one such color.

There are a host of studies that show the color pink has a persuasive effect. For example, door-to-door charity workers who wore pink shirts saw donations rise threefold. Schools who painted their walls pink noted that students were calmer and more engaged. Research has also revealed the color pink suppresses angry behavior among prisoners. The color pink seems to have a tranquilizing effect.

13. Lead Others to Self-Discovery

When you're trying to persuade people to do something, such as convincing a reluctant employee to change a negative behavior, don't tell them why they should change. Instead, help people discover their own reasons for changing. A Yale School of Medicine study found people usually act for their own reasons, not someone else's.

The process involves asking the person you're trying to persuade six questions, two of which are:

1. How ready are you to change on a scale from 1 to 10, where 1 means "not ready at all" and 10 means "totally ready"?

2. Why didn't you pick a lower number? Or if the person picked a 1, ask them what it would take to turn that 1 into a 2.

This second question gets people to articulate their own reasons for doing something. When people do this, they're more likely to buy into those reasons and make a stronger effort to orchestrate the change.

14. Repetition is the Key

Skepticism is at an all-time high among consumers. Research has shown the majority of people need to hear company information three to five times to believe messages. Keep this in mind when you set out to deliver a message with the goal to persuade. Make a point of repeating your message at different intervals, through different mediums. Familiarity, it seems, leads to acceptance.

15. The Right Messenger

Choosing the right messenger is a very effective persuasion tactic. A messenger who shares common values and perspectives with the audience typically has a greater probability of success in persuading them , even if the specific point or objective of the message is not initially favored or being considered. Such a person can more easily establish common ground and be less threatening to the audience's core values.

16. Use Monroe's Motivated Sequence

The late Alan H. Monroe of Purdue University has a time-tested organizing structure proven to persuade people. His five-step process is summarized below:

Step 1: Capture the attention. Hook your listeners right away and make them want to be attentive to your message. There are several gambits to achieve this, such as a brief story, a startling statistic, a powerful quote, a remarkable visual, a question or a rhetorical question.

Step 2: Establish the need. Explain why there's a problem needing to be addressed. This is the step that makes your audience realize why they should care. You want them to think "I need to hear this" or "We must do something about this!"

Step 3: Satisfy the need. Tell your listeners how you're uniquely positioned to fill that need. This is your solution to the problem or issue that requires attention.

Step 4: Visualize the results. Take your listeners through a positive visualization, that is, help them see the benefits of adopting your solution or proposal. Then take them through a negative visualization—help them see the downsides of not taking any action. Your aim here is for them to think, "This is a great idea."

Step 5: Ask for the action. Clearly outline the action you want them to take. The goal here is for them to decide "I want to do this. Show me how to get to it."

17. Use Baby Pictures

Our brains interpret baby images as powerful messages. They can also influence us. Baby images add an emotional component that increases the persuasive power of the message.

In *Brainfluence: 100 Ways to Persuade and Convince Consumers with Neuromarketing*, author Roger Dooley studied how people view ads with babies in them. Employing eye-tracking technology to measure the direction and duration of a viewer's eye movement, researchers discovered viewers fixated on baby faces and gave a significant less amount of attention to headlines and ad copy. But when a side-facing baby was looking toward the ad's headline, the ad headline and copy got far more attention than when the baby was not looking at the headline.

18. Last Chance

According to Robert Cialdini, Ph.D., a world authority on persuasion, eliminating the risk of losing something is more attractive to someone than increasing the prospect of winning something.

Loss language is motivating, Cialdini says, because it taps into the "scarcity principle." As discussed earlier in the chapter, the scarcity is principle is an effective means of persuasion. It works on the premise that what is scarce is more desirable.

The Keys to Influence

To be an effective persuader, you cannot use the same techniques for all people all the time. You must customize your message to fit the demographics, interests, and values of your audience.

The first step of persuasion is to be mentally ready to persuade. An important part of preparing yourself is to knowas much about your audience as possible.

One you understand your audience, you need to consider which persuasion methods will be most effective and plan the persuasive techniques you will used instead of "flying by the seat of your pants.'" Think of the preparations you make before traveling such as looking at a road map and weather conditions before. . You need to understand where you are going, what route you should take, and what the driving conditions will be. Persuasion operates the same way. Use the three D's principle: discover, design, and deliver:

- **Discover** what your prospects want and need to hear.
- **Design** and structure a winning persuasive argument.
- **Deliver** the message with passion, compassion, and purpose.

The key is to understand the situation or the prospect in order to implement the most effective methods of persuading your audience.

The Mental Game of Persuasion

Beliefs

Understanding your audience's beliefs will help you know what approach to take. Beliefs are those things we accept as truth, consciously or subconsciously, proven or unproven. Beliefs come from our environment, our culture, our education, our experience, and even through osmosis from our friends and family. One of the most common sources of our beliefs comes from being a part of a group, such as a family or a tight-knit community. People often take on the beliefs and rules of the groups to which they belong and then behave in accordance with those beliefs and rules.

Values

A value is more ingrained than a belief, because people are more deeply and consciously committed to their values. A value is typically something that has been very thoroughly contemplated and accepted. It is for this reason that values are much harder to change than beliefs. Usually, a true value will not be changed—even by wealth, acceptance, or pressure. Be sure when your audience never feels as though you're trying to attack their values, when you are attempting to persuade them to do something. This will only make them feel defensive toward you. As Walt Disney wisely stated, "When values are clear, decisions are easy."

Indifference

People who are indifferent most likely have never even thought about the issue, or they have previously had no reason to care about it. Indifferent people come across as greatly apathetic, because the topic you are presenting is something they've never had to cognitively process before. People who are indifferent don't want to be bothered. These people usually don't care about you or your message and are often only there, because they have to be. Sometimes their indifference is just a general lack of interest or boredom in general. An indifferent audience needs attention, empathy, and a reason to care.

Determine Where the Audience Stands

As discussed above, an important part of persuasion is determining what the audience's current acceptance level is regarding the subject you want to present. Ask yourself the following questions when making this determination:

1. **Knowledge**: What does my audience know about the topic?
2. **Interest**: How interested is the audience in my subject?
3. **Background**: What are the common demographics of my audience?
4. **Support**: How much support already exists for my views?
5. **Beliefs**: What are my audience's common beliefs?

Understanding different types of audiences will also help you determine their acceptance level. Following are some different categories of audiences and how to deal with each main type.

A Hostile Audience

A hostile audience disagrees with you and may actively work against you. For this group, use the following techniques:
* Find common beliefs and values.
* Use humor to break the ice.
* Don't start the presentation with an attack on their position.

- Don't talk about anything else that could be considered hostile. You are only trying to persuade on one point.
- Increase your credibility with studies from experts or anything that supports your claim.
- Don't give them reasons to not like you.
- Don't tell them you are going to try to persuade them.
- Express that you are looking for a win-win outcome rather than a win-lose situation.
- If possible, meet with the audience more than once before confronting them on areas of disagreement.
- Show them you've done your homework.
- Respect their feelings, values, and integrity.
- Use logical reasoning as clearly and carefully as possible.
- Use the Rule of Connectivity and the Rule of Balance.

A Neutral or Indifferent Audience

This audience understands your position but isn't necessarily interested in the outcome. The key to dealing with a neutral or indifferent audience is creating motivation and energy. Be dynamic, energetic, and charismatic while implementing one or more of the following techniques:

- Spell out the benefits to them or the things around them.
- Point out the downside of not accepting your proposals.
- Grab their attention by using a story. Make them care by showing them how the topic affects them.
- Get them to feel connected to your issues.
- Avoid complex arguments.
- Use concrete examples with familiar situations or events.
- Identify why they should care.
- Use the Rule of Involvement and the Rule of Social Validation.

An Uninformed Audience

An uninformed audience lacks the information they need to be convinced. To persuade them, use one or more of the following tactics:

- Encourage them to ask questions throughout the presentation.
- Keep the facts simple and straightforward.
- Find out why they are uninformed.
- Use examples and simple statistics.
- Quote experts the audience respects.
- Stress your credibility, such as degrees, special expertise, and experience.
- Make your message interesting in order to keep their attention.
- Use the Rule of Dissonance and the Rule of Scarcity.

A Supportive Audience

A supportive audience is one that already agrees with you. You may think persuading these people will be easy; however, there is a difference between agreeing with you and actually doing something. Your goal is to get them to take action. The following techniques should be used with a supportive audience:

- Increase energy and enthusiasm with inspiration.
- Prepare them for future attacks by inoculating them against other arguments.
- Get them to take action and to support your cause.
- Let them know what needs to be done.
- Use testimonials to intensify the commitment.
- Use the Rule of Esteem and the Rule of Expectation.

Most audiences are a mix of the types of audiences described above. Determine the dominant audience type with which you will be dealing and tailor your remarks accordingly, while mixing in some techniques for the other audience types.

When Persuasion Backfires

As mentioned above, understanding your audience and applying the appropriate techniques of persuasion provides the highest probability of success. However, there are times and situations when certain persuasive techniques are not appropriate. We cannot treat every person or every audience the same way. If we take persuasion too far, we will run into problems.

People can be persuaded and influenced, until they feel cheated, misled, or taken advantage of, and then they develop adverse feelings that result in negative outcomes which may include never doing business with you again.

Some sales and marketing campaigns have a tendency to push the envelope a little too hard when trying to persuade others. Persuaders who do not possess the ability to read others or who do not have the skills necessary to persuade typically fall victim to persuasion backfire. This typically happens when they take persuasion too far, using extreme pressure or trying to sell a product that is neither needed nor wanted. Use persuasion, influence, or power the wrong way and people lose all trust in you, eliminating the likelihood you will ever be able to persuade them again. Over-persuasion sets off silent alarms in you prospects' minds. It could be a feeling of uneasiness, or a bad feeling toward you, your service, your product or your business.

This backfire also includes selling a faulty product or providing poor service. Often people will say nothing to you about the defective item, poor service, or your pushy tactics. They will simply never do business with you again or want to associate with your product, service or business. They may even give you negative reviews online. They will never trust or listen to your point of view again. This pitfall is a silent killer, because most persuaders don't even realize the mistake was ever made. You have probably had this happen to you and can relate to this negative feeling towards a sales person or business.

No one likes feeling manipulated or pressured. We have all been burned or taken advantage of, and when we see signs of such behavior we put up walls or go elsewhere. Uneducated persuaders can be offensive, condescending, obnoxious, and insulting. We all have a conscience and tend to have a sixth sense when it comes to persuasion as to how hard we can push our view. We must obey our inner voice when it nudges us to use restraint and/or try a better persuasive tactic for the listener(s) or situation. Effective persuasion means becoming more knowledgeable, refining our presentation via practicing good techniques, and reading our listeners.

Fine-Tuning Our Persuasion Radar

The more we understand human nature and personality types, the better we will be able to customize our persuasive presentations. We run into trouble when we try to put people in a box or categorize them, but the reality is (most of the time) people are predictable. Sure, people can never be 100 percent predictable, but it is amazing how predictable people actually are as one studies human nature.

Polished persuaders understand personality types should dictate how they customize their message. When you analyze personality types, ask yourself the following four questions:

1. Is the audience or prospect mostly logical or emotional?

Logical Types

- Use their heads
- Go with what makes sense
- Use facts, figures, and statistics
- Rely on past history

Emotional

- Use their heart
- Go with what feels right
- Use emotional appeals
- Rely on intuition (their 'sixth sense')

2. Is the audience or prospect mostly introverted or extroverted?

Extroverts

- Love to communicate
- Are talkative
- Involve others
- Tend to be public people
- Desire face-to-face contact

Introverts

- Keep feelings inside
- Listen more than they talk
- Like to work solo
- Tend to be private
- Prefer memos and e-mails

3. Is the audience or prospect motivated more by desperation or inspiration?

Desperation

- They prefer to get away from the problem
- Often get stuck in the past, don't want to repeat prior mistakes
- Strongly want to avoid pain and discomfort
- Desire to get away from something

Inspiration

- Work towards a solution
- See a better future
- Are motivated by pleasure
- Want to move forward (have a vision)

4. Is the audience or prospect assertive or amiable?

Assertive

- Consider results more important than relationships
- Make decisions quickly
- Want to be in control
- Are task-oriented
- Don't waste time
- Are independent

Amiable

- Consider relationships more important than results
- Are friendly and loyal
- Like to build relationships
- Are great listeners
- Avoid contention
- Are nonassertive and agreeable

Understanding your audience or prospect and adapting your persuasive message to their personality type will enable you to have much greater success.

Structuring a Successful Message

There are several basic elements to any persuasive message. The personality type, situation, and communication method should dictate how much time and effort is invested in these elements.

For maximum probability of success the persuasive message should focus on one defined issue. The more persuasion objectives you try to achieve in each message the lower the success rate. Therefore, stay focused on one specific objective and refrain from any issues that do not coincide with the objective. The structure of your persuasive message should follow the pattern discussed below.

1. Create Interest

You have to generate an interest about your chosen topic. Your audience needs a reason to listen: Why should they care? What's in it for them? How can you help them? A message that starts with a really good reason to listen will grab the attention of the audience, enabling you to continue with the message. Without their attention, there is no hope of getting your message across.

2. State the Problem

You must clearly define the problem. The best pattern for a persuasive speech is to find a problem and relate how it affects the audience. In this way, you show them a problem they have, and why it is of concern to them. Why is this a problem to your audience? How does this problem affect them?

3. Offer Evidence

This is the support you give to your message. Evidence validates your claims and offers proof your message is right for them and allows your audience to rely on other sources in addition to you. Evidence can include examples, statistics, stories, testimonies, analogies, and any other supporting material that enhances the integrity and congruency of your message.

4. Present a Solution

You have gained your audience's interest and provided evidence in support of your message; now, you must solve their problem. You present the message you want them to believe and satisfy the need you have identified or created. You have created dissonance and now you are providing the solution. For instance, express how your product or service meets their needs and helps them achieve their goals.

5. Call to Action

A persuasive message is not truly persuasive, if your audience does not know exactly what they need to do. Be specific and precise. In order to complete the solution to their problem, they must take action. This is the climax, the peak of your logical and/or emotional appeal. The prescribed actions must be feasible. Lastly, do everything possible to make your call to action as easy as possible. For instance, provide them with an order form already filled out.

Using this type of structure facilitates people's acceptance of your message and clarifies what you want them to do.

In order to create a good structure for your argument and to reach your audience, it may be helpful to consider the following set of questions.

Ask yourself these questions in regard to yourself and your message:

- What do I want to accomplish? (What is my objective?)
- What will make my message clear to my audience?
- What will increase my credibility and trust?
- Which persuasion techniques will work best for my audience?
- What do I want my prospects to do?

Ask yourself these questions in regard to your audience:

- Who is listening to my message? (Audience demographics)
- What is their initial mindset? (What are they thinking and feeling now?)
- When will the call to action work? (What do you want them to do and when do you want them to do it?)
- Why should they care? (What is in it for them?)
- In what areas of their lives does this affect them? (Health, money, relationships, etc.)
- How will they benefit? (What will they gain?)

These questions will help you create effective arguments in each of the key areas: interest, problem, evidence, solution, and action.

Change: Opening Up a Closed Mind

Life is change; persuasion is change. You must be able to create and motivate change. It is human nature to resist change and hold onto comfort zones as people tend to follow the path of least resistance. However, change is what helps us learn and grow. We all want to become a better person and be "stretched" to accomplish more things, but we so easily get stuck in our daily patterns.

Consider how resistant to change your audience is likely to be. Will persuading them be like breaking through a brick wall or a cardboard box? Are they ready to make changes because of their circumstances and surroundings? Are they already trying to change? Don't get discouraged if some of your listeners oppose you and resist your persuasive message.

There are three main ways people make changes in their life. One is through drastic change. This could be a heart attack, a personal tragedy, or losing a job. These events force people to change their lives. They did not feel a need to change until threatening, life-changing events occurred. The second is through gradual change. This is a process that evolves from events or personal relationships. Gradual change happens over time, and sometimes it is such a slow process it is difficult to notice it is happening at all.

The third way people change their lives is through internal change. This can come from inspiration or desperation, but either way, they have consciously decided they are going to make changes in their life.

To get change to stick, you must make sure three things occur: First, there must be a long-term, enthusiastic commitment to change. People have to decide there is no other option. The second is they must be willing to pay the price (i.e. discipline, disruption of a routine, etc.) and remain persistent even when feeling weak. Third, people have to have a clear picture of the end result of the change. How is this going to affect their life? What is the outcome?

The biggest obstacles to change are lack of motivation, lack of knowledge, and fear. People will not change if they don't understand the negative consequences of not changing or the positive rewards of making the change.

Also, realize that people will resist change unless sufficient reinforcement and tools are provided to assist them. Without having the proper knowledge, resources, and assistance lasting change will likely not be successful.

As a persuader, you need to create a vision for your audience; one that shows them what they will be like in the future. If you can get people to see themselves in the future and understand where the change will take them, they will be more willing to embrace the change.

A Call to Action

A 'call to action' was briefly addressed to in a few places earlier in the chapter. This section will discuss it in more detail, as the call to action is the most important part of your presentation. This is where you communicate exactly what you want your listener(s) to do.

This conclusion should not come as a shock to your listener(s). Throughout your presentation, you should have gently led them to the same conclusion you are now giving them. You should have already prompted them to want to do what you are about to tell them to do.

Many people struggle with this part of persuasion, because they are asking their prospects to do something. However, this should be the best part — the action is the objective of the presentation. If you become tense and uneasy, so will your listener; it is imperative to remain confident and pleasant during the call to action phase. The whole presentation should be structured to make the call to action smooth and seamless. In fact, the audience should not even see or feel your transition into your call to action.

Your entire presentation should be built around or focused on the call to action. When planning and preparing your call to action, remember the process does not have to be long and painful; it can be short, brief, and to the point. If possible, write out the call to action word for word beforehand. From the outset of your message, you must be eager to get to this point and finish strong. Be positive and enthusiastic. Regardless of how you say it, you need to lead them to a belief or provide specific instructions on what to do. Make the call to action easy for them to follow and simple for them to do and be crystal clear about it; there should be no doubt in your listeners' minds exactly what you want them to do.

Structuring the Call to Action Message

Once the call to action has taken place, your audience needs to remember, retain, and respond to your message. They have to keep doing what you want them to do. Your responsibility is to ensure your points are memorable, easy to understand, and simple to follow. The following critical items must be included in your persuasive presentation.

1. Repetition

The use of repetition is very effective. Research has shown repetition plays an important part of learning; it is also important for effective persuasion. Repetition creates familiarity toward your ideas, and that leads to a positive association. When something gets repeated, it gets stuck in our memory and improves our comprehension.

Make it a point to repeat your message several times, so your audience understands precisely what you are talking about and comprehends exactly what you want them to do. You can repeat the message several times without saying the same thing over and over again by repackaging how you say it—use new evidence and new words, so you don't sound like a broken record. Using repetition too much runs the risk of diminishing returns. You know how you feel about someone telling you a joke or a story you've already heard or about that commercial you've seen one too many times. If your audience hears the same thing too many times, they will tune you out and quit listening.

2. Theme

We see general themes in commercials and advertisements. A theme is easily remembered and easily retained. Having a theme will give your presentation flow, order, and presence in the minds of your audience. Themes provide an easy way for people to remember the main point of your message. If you have strong and well-organized themes, you can be sure your audience will understand and remember your message more clearly and more strongly.

3. Brevity and Simplicity

Keep your message short and simple. Boring an audience has never worked as an effective persuasive technique. If the message is short and simple, it will most likely be clearer and therefore easier to remember. Consider the profundity of Abraham Lincoln's historical Gettysburg Address. The whole speech, from start to finish, was only 269 words. He presented it in less than three minutes.

Winston Churchill read his "blood, sweat, and tears" speech in less than two and a half minutes. Even

Nelson Mandela's famous speech signaling the end of apartheid — a speech he gave after twenty-seven years of imprisonment — lasted only five minutes.

Make sure your speech is articulate and intelligent but avoid using esoteric language. Use simple terms and jargon familiar to your audience. Complexity will not impress the audience; it will only muddle the message. Make your points simple, clear, and direct and avoid referencing too many facts, figures, examples, questions, or anything else that might complicate your message.

4. Timing the Message

Timing can be everything. Time your message so it will have its greatest impact. The impact points should be made at the beginning and end of a presentation. These effects can be powerful presentation tools as the first and last parts of your presentation typcially have the most impact on your audience's overall impression. Your first and final words determine how you will be remembered and thought of long after your speech has ended. With that in mind, carefully craft your opening and closing statements, placing your strongest points at those junctures.

5. Offer Choices

This option is somewhat controversial among persuasion experts. Many believe persuasive messages should conclude with only one option; however, there are also many in the persuasion circles who believe offering multiple options is the best approach. This section is provided with the latter in mind. The author is of the opinion that there are situations when one option is best and other instances when offering multiple options is a better path to obtaining the desired outcome.

There is a strange, psychological phenomenon in regard to drawing conclusions. If someone tells us exactly what to do, our tendency is to reject that dictated choice. Offering your prospects a few options, allowing them to make the choice for themselves is a good solution. People like having the have freedom to make their own choices. If forced to choose something against their will, they experience psychological resistance and feel a need to restore their freedom.

The strategy is to only provide options that will satisfy your situation and objective. In sales, they call this strategy the alternative close. For example, have you heard the line, "Do you want regular or deluxe?" Or what about, ""Do you want to meet Monday afternoon or Tuesday evening?" The person has options, but both options meet the persuader's goals.

While options are good for your audience, avoid giving them more than two or three choices. If you give too many alternatives, your prospect(s) will be less likely to choose any of them. Structured choices give the audience the impression of control. As a result, they increase cooperation and commitment.

Each option offered meets the persuader's objective without making him appear as if he is restricting freedoms. Even if it is something simple, people desire to have options. By simply using the word "or" you provide a call to action closing that empowers the listener to make a decision that meets your objective.

Prepping the Opposition

When you are presenting, and you realize there is an opposing viewpoint, it is helpful to prepare your audience in advance. The idea is to address the issues your opponent will bring up and directly refute them.

We are surrounded by countless examples of this successful approach. For instance, the attorney in the courtroom stands up and says, "The prosecution will call my client mean, evil, a terrible husband, and a poor member of society, but this is not true, as I will show you over the next couple of weeks.. .." So, when the prosecutor stands up and states anything close to what the defense attorney has claimed she will, the jury is prepared, thinking she is acting exactly the way the defense said she would. This gives the jurors a way to ignore or even discount the prosecutor's arguments.

Street gangs also use this method. When they are attempting to convert someone to their beliefs and to join the gang, they will prepare the future gang member by telling him his parents, teachers, and cops will encourage him not to join a gang. They will tell him all the reasons his opponents will give, fueling him with ammunition for the impending attack. This preparation enables him to handle the oncoming assault from parents, teachers, etc.

When you prep people in advance they mentally prepare arguments supporting their stance. This reinforcement prevents them from switching teams. The more prepared they are, the more they'll hold fast to their attitudes and beliefs. The more deeply this reinforcement is ingrained, the more difficult it will become for them to be swayed later.

If the listener(s) already agree with your position, prepping is not necessary. If they disagree with you, however, there are persuasive advantages to presenting both sides of the argument. This persuasion technique is especially effective if an opposing speaker will be presenting after you. Giving both sides of the argument also works better with audience members who already know something about the opposition's strength.

By presenting the audience with the other side of the argument, you show the audience you are not afraid of the truth and have done your research. You prepare your audience in advance about the negative things someone may say about you, your service or product. You will win a great deal of respect and power when you answer someone's questions, before they even ask them.

Preparation Is the Key to Influence

Persuasion is everything. Understanding the foundation and principles of persuasion will help you be a more effective communicator and bolster your professional and personal success. The underlying foundation for using any persuasion tools, however, needs to be authenticity and integrity—these are important pillars for earning trust.

Prepare your mind, know your audience and structure a winning persuasive argument accordingly. Effective persuaders learn about their audience and implement the best persuasion techniques for each situation.

CHAPTER 13

Sales & Selling: Being Successful

The Importance of Sales in an Organization

In any business organization, sales is the department that generates revenue. No matter how good the product or service, how cutting-edge the technology, or how progressive and forward-thinking the company, without sales everything else is useless.

Sources of Revenue

A business organization can generate revenue from a variety of sources including operating income from sales, royalties, dividends and interest; income from financial assets it owns; payouts from insurance policies, rental income; and capital gains from the sale of owned properties. However, even when organizations derive little or no taxable income directly from operations stemming from sales—such as an investment company—they must still generally have some sort of sales effort to generate investment revenue to fund their investments in income-producing assets. For example, a limited partnership must often engage in a concentrated sales effort to recruit more limited partners.

Sales versus Marketing

While it is sometimes difficult to draw the line where the marketing process ends and the sales effort begins, the sales effort is the effort that actually collects money—or the obligation to buy, in the case of a purchase order or finance arrangement. The marketing effort creates favorable conditions for the sale to take place. In a nutshell, the marketer leads the horse to water; the sales team makes it drink.

Partnership between the Sales and Marketing Teams

Few products are truly bought on impulse. Even a can of brand-named soda on a shelf is there, because its wholesaler built a relationship with the store manager

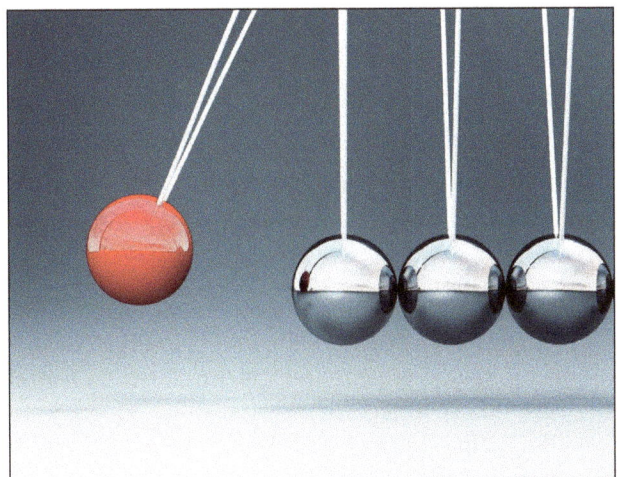

over time and secured good shelf placement, denying it to a competitor. To get the most out of the sales effort, the sales team needs support from the marketing team to facilitate follow-up contacts, mailings and account service. If the support is not there, the account may not last long, and turnover will increase. If the sales staff is too directly involved in that effort, it may eventually become overwhelmed with account services and find it difficult to grow the business. Sales is so important, then, that it typically behooves management to free its sales staff from some or all of the account services process to generate future revenue.

Investing In Sales

Many companies under-invest in their sales effort, treating sales like an afterthought to be handled after the managers solve all the manufacturing, distributing and financing issues. The best sales forces are built with professional, well-compensated employees. They aresupported by a strong marketing effort and empowered to act, serving key client interests with marketing support, money and time. They have strong personal relationships with key customers, or they learn how to build them.

Big-Ticket Necessity

Sales is more common or more intensive when companies market big-ticket items, such as cars, appliances and furniture. Large purchases usually require more persuasive efforts and communication with prospects. Companies often use advertising and other promotions to build awareness and attract potential customers. Sales reps then either go out to meet with prospects or greet them when they come in looking to buy. Because of the risks of bigger purchases, buyers usually need a more thorough explanation of benefits and the value proposition.

Sales Conversions

A main role of sales is to improve the efficiency of converting prospects into customers. Salespeople can directly interact with prospects, ask questions to help, address any possible buyer concerns and ultimately recommend products or services. Without these important sales steps, a company has to rely on passive marketing messages to carry weight, when buyers are looking to make a purchase decision. Advertising, for instance, can't answer follow-up questions customers have after being exposed to the original ad messages. Salespeople have the benefit of back-and-forth discussion.

Growth

The sales function is a key mechanism for companies to grow through referrals. In fact, a number of sales organizations indicate in collateral materials or thank-you letters that referrals are essential to their businesses. Leveraging the fact that a new or established customer sees value in the solution, a sales rep can ask if the customer knows other people with similar needs or interests. Referrals are an efficient prospecting tool, because they create a personal connection to the new prospect.

Customer Retention

The personal, interactive nature of selling makes it a key ingredient in company efforts to build long-term relationships with customers. Salespeople can follow up after purchases to ensure customers have a good experience. Without this contact, upset customers often don't complain; they just go away to other providers. This ongoing interaction also allows for opportunities to make additional sales that address future or ongoing needs of the customer.

10 Reasons Why Top Salespeople are Successful

Brian Tracy is Chairman and CEO of Brian Tracy International, a company specializing in the training and development of individuals and organizations. He has consulted for more than 1,000 companies and addressed more than 5,000,000 people in 5,000 talks and seminars throughout the US, Canada and 70 other countries worldwide. As a keynote speaker and seminar leader, he addresses more than 250,000 people each year.

Brian Tracy's research has shown that the top 20 percent of salespeople earn 80 percent of the money. His studies have shown that the top salespeople have integrated the following vital keys for success in sales.

Key to Success #1: Top Salespeople Do What They Love to Do

All truly successful, highly paid salespeople, love their sales careers.

You must learn to love your work and then commit yourself to becoming excellent in your field.

Invest whatever amount of time is necessary to improve your sales career — pay any price; go any distance; make any sacrifice to become the very best at what you do. Join the top 10 percent.

Key to Success #2: They Decide Exactly What They Want

Don't be wishy-washy. Decide exactly what it is you want in life. Set it as a goal for your sales career and determine what price you are willing to pay to get it.

According to the research, only about 3 percent of adults have written goals, and these are the most successful, highest-paid people in every field. They are the mover and shakers, the creators and innovators, the top salespeople and entrepreneurs.

Key to Success #3: They Back Their Sales Career Goals with Perseverance

A **key to success** in sales is to back your goal with perseverance and indomitable willpower. Decide to throw your whole heart and soul into your success and into achieving your **sales career** goal. Make a complete commitment to improve your sales career and become one of the most highly paid salespeople. Resolve that nothing will stop you or discourage you.

Key to Success #4: They Commit to Lifelong Learning

Your mind is your most precious asset, and the quality of your thinking determines the quality of your sales career.

Commit yourself to lifelong learning.

I cannot emphasize this too often. Read, listen to audio programs, attend seminars, and never forget that the most valuable asset you will ever have is your mind. As you continue to learn, you will eventually become one of the most valuable salespeople in your company.

The more knowledge you acquire that applies to practical purposes, the greater will be your rewards, and the more you will be paid.

Key to Success #5: Top Salespeople Use Their Time Well

Your time is all you have to sell. It is your primary asset. How you use your time determines your standard of living.

Resolve, therefore, to use your time well. Begin every day with a list. The best time to make up your work list is the night before, prior to wrapping up for the day. Write down everything you have to do the next day starting with your fixed appointments and moving on to everything else you can think of.

Key to Success #6: They Follow the Leaders

Do what successful people do. Follow the leaders, not the followers. Do what the top salespeople in your company do. Imitate the ones who are going somewhere with their lives. Identify the very best salespeople in your field and pattern yourself after them.

If you want to become one of the best salespeople in your company, go to the top earners and ask them for advice. Ask them what you should do to improve your sales career. Inquire about their attitudes, philosophies, and approaches to work and customers.

Key to Success #7: They Know That Character is Everything

Guard your integrity as a sacred thing. Nothing is more important to the quality of your life in our society. In business and sales success, you must have credibility.

You can only be successful if people trust you and believe in you.

In study after study, the element of trust has been identified as the most important distinguishing factor between one salesperson and another and one company and another.

Key to Success #8: They Use Their Inborn Creativity

Think of yourself as a highly intelligent person, even a genius. Recognize that you have great reserves of creativity you have never used.

Say aloud, over and over, *"I'm a genius! I'm a genius! I'm a genius!"*

This may sound like an exaggeration, but it isn't. The fact is that every person has the ability to perform at genius levels in one or more areas. You have within you, right now, the ability to do more and be more than you ever have before.

Key to Success #9: They Practice the Golden Rule

Practice the Golden Rule in all your interactions with others: Do unto others as you would have them do unto you.

Think about yourself as a customer. How would you like to be treated?

Obviously you would want **salespeople** to be straightforward with you. You would want them to take the time to thoroughly understand your problem or need and then show you, step by step, how their solution could help you improve your life or work in a cost-effective way.

If this is what you would want from a salesperson selling to you, then be sure to give this to every customer to whom you talk.

Key to Success #10: They Pay the Price of Success

Finally, and perhaps more important than anything else, resolve to work hard. This is a great key to success in life.

The key to success in selling is for you to start a little earlier, work a little harder, and stay a little later. Do the little things that average people always try to avoid doing. When you begin your workday, resolve to *"work all the time you work."*

The 10 Laws of Sales Success

Entrepreneur is a North American magazine and website that carries news stories about entrepreneurship, small business management, and business. The magazine was first published in 1977, and provides practical advice on entrepreneurship and small business.

Entrepreneur emphasizes that selling can be one of the most rewarding tasks you'll undertake as a business owner, if you follow these 10 tactics:

Law #1: Keep your mouth shut and your ears open.

This is crucial in the first few minutes of any sales interaction. Remember:

Don't talk about yourself.

Don't talk about your products or services.

And above all, don't recite your sales pitch!

Obviously, you want to introduce yourself. You want to tell your prospect your name and the purpose of your visit (or phone call), but what you don't want to do is ramble on about your product or service. After all, at this point, what could you possibly talk about? You have no idea if what you're offering is of any use to your prospect.

Law #2: Sell with questions, not answers.

Remember this: Nobody cares how great you are until they understand how great you think they are. Forget about trying to "sell" your product or service and focus instead on why your prospect wants to buy. To do this, you need to get fascinated with your prospect; you need to ask questions (lots of them).

Law #3: Pretend you're on a first date with your prospect.

Get curious about them. Ask about the products and services they're already using. Are they happy? Is what they're using now too expensive, not reliable enough, too slow? Find out what they really want. Remember, you're not conducting an impersonal survey here, so don't ask questions just for the sake of asking them. Instead, ask questions that will provide you with information about what your customers really need.

When you learn what your customers need, and you stop trying to convince or persuade them to do something they may not want to do, you'll find them trusting you as a valued advisor and wanting to do more business with you as a result.

Law #4: Speak to your prospect just as you speak to a well-respected friend.

There's never any time you should switch into "sales mode" with persuasion clichés and tag lines. Affected speech patterns, exaggerated tones, and slow, hypnotic sounding "sales inductions" are never acceptable in today's professional selling environments. Speak with friendly enthusiasm, but keep it real.

Also, leave out the profanity. Profanity is offensive and makes one come across as being disrespectful. Many also hold the opinion that those using profanity lack the intelligence and discipline to use a better choice of words.

Law #5: Pay close attention to what your prospect isn't saying.

Is your prospect rushed? Does he or she seem agitated or upset? If so, ask if there is anything you can do to be of assistance, and if you can schedule a better time to meet. Some salespeople are so concerned with what they're going to say next that they forget there's another human being involved in the conversation.

Law #6: If you're asked a question, answer it briefly and then move on.

Be respectful of their time and keep the conversation focused on meeting their needs.

Law #7: Only after you've correctly assessed the needs of your prospect do you mention anything about what you're offering.

Know with whom you're speaking and focus on identifying their needs.

Law #8: Refrain from delivering a three-hour product seminar.

Don't ramble on and on about things that have no bearing on anything your prospect has said. Pick a handful of things you think could help with your prospect's particular situation and tell him or her about them. (If possible, reiterate the benefits in his own words, not yours.)

Law #9: Ask the prospect if there are any barriers to them taking the next logical step.

After having gone through the first eight steps, you should have a good understanding of your prospect's needs in relation to your product or service. Knowing this, and having established a mutual feeling of trust and rapport, you're now ready to bridge the gap between your prospect's needs and what it is you're offering.

Law #10: Invite your prospect to take some kind of action.

This principle obliterates the need for any "closing techniques" because the ball is placed in the prospect's court. A sales 'close' keeps the ball in your court and all the focus on you, the salesperson. You don't want the focus on you. You don't want the prospect to be reminded that he or she is dealing with a "salesperson." You're not a salesperson; you're a human being offering a particular product or service to fulfill their need. If you can get your prospect to understand this fact, you're well on your way to becoming an outstanding salesperson.

Improve Your Sales Conversations

If you want to improve your sales conversations, pay attention to these 7 keys:

1. **Build rapport**: Before you ask questions to get the buyer to open up or talk about how you can help, you have to build rapport. All else being equal, people buy from people they like. Be likable and focus on relationship building, and you'll find your sales conversations will go much more smoothly.

2. **Uncover aspirations and afflictions.** If you've ever read any piece of sales advice, you know you need to ask questions to uncover the prospect's pain. That's a given. What most advice doesn't include is how to harness the power of aspirations. Your job is not only to uncover the prospect's needs and pains but to also uncover their aspirations and goals. Get your prospect to open up and share his/her hopes, dreams, and desires and then demonstrate how you can help achieve his/her goals.

3. **Make the impact clear.** If you don't make the business case, you won't make the sale. You can do everything else right, but if the prospect doesn't see the value of your solution (and you need to be very clear with what that value is), they will not buy it.

4. **Paint a picture of the new reality.** This goes hand in hand with points 2 and 3. Once you know the prospect's needs and goals and the tangible impact of alleviating these pains or attaining his/her goals, you must paint a picture of what his/her new world will look like. How will it be better? In your sales conversations, help visualize the other side and build excitement around it.

5. **Balance advocacy and inquiry.** Sales conversations require give and take. You have to get the prospect talking so you can fully understand his/her situation. You also need to take what the

prospect says and communicate recommendations based on your expertise to help him/her see how you can help. In each and every sales conversation (yes, this includes capabilities presentations and demos) you have to balance how much you talk and how much you listen.

6. **Build on the foundation of trust**. Trust is the foundation of sales success. A buyer will not open up and share his/her needs if he/she doesn't trust you. A buyer will not believe in your solution and that you can do what you say you can do, if he/she doesn't trust you. If there is no trust, a buyer will never see the full value of what you propose, and you will not win the sale.

7. **Plan to succeed**. Set the table for success by going into each sales conversation with a plan. Do your homework and know what you want to get out of the conversation. If you go into each conversation well-prepared and planning to succeed, you will be much more likely to make the sale.

If you follow these seven keys in your sales conversations, will you still make mistakes? Absolutely. Will you win every sale? Absolutely not. After all, we're only human, and no human is perfect.

But, these keys will help you avoid the common mistakes many sellers make and help you lead more successful and productive sales conversations.

Seven Ideas for Building Trust in Sales

Many things have changed in the world of sales, but some have not. Building trust was important 50 years ago, and it's just as important today. When buyers trust sellers, they depend on them, listen to them, give them access, and spend time with them.

Trust is critical for sales success, but today's buyers are busier than ever, and, at the same time, have access to more information and choices. This makes their time harder to get, and their trust harder to build.

Building trust is one of six key drivers of client loyalty and one of the top 10 things sales winners do. Trust in sales is built around three factors—competence, integrity, and intimacy.

Competence

Trust in a seller's competence means buyers believe you can do what you say you can. This is not referring to trust in the product or service, but trust in the seller as a person. Buyers need you to bring ideas to the table, to help them find solutions to problems, and give sound advice. If they don't trust your competence, they won't accept the advice.

Demonstrating your competence in the following three ways will go a long way towards building trust:

1. **Be an expert.** Too many buyers report they don't trust sellers, because the sellers don't know their stuff. As a seller you need to know your buyers' industries and businesses, competition, marketplace, full set of customer needs, and more—inside and out. You must answer buyers' questions about your offerings and the market,

as well as about the buying process itself. If you want to guide the way, you need to make it your business to be a source of knowledge in all of these areas.

2. **Know your impact model.** Sales winners craft compelling solutions. The key to making solutions compelling is a concrete return on investment case. Be prepared to discuss, in concrete terms, what results buyers can expect to achieve. If you don't know the impact model—how you can affect their business—buyers will not trust your business sense.

3. **Develop and share a point of view.** If a buyer is confused about what to do to resolve their problem, and you don't have an answer or are unwilling to develop and share a point of view, they won't see you as a trusted advisor. Part and parcel of being an advisor is you know your stuff and the buyer wants and values your opinion.

Integrity

Everyone has been sold a bill of goods and not gotten what they were promised. You may think your integrity is off the charts, but buyers have been burned before and are suspicious before ever meeting you. You must demonstrate and prove your integrity. Buyers won't just assume it is there. To demonstrate and prove inegrity:

1. **Demonstrate moral principles.** Successful sellers always do the right thing, even in morally ambiguous situations. This can mean turning down business, suggesting alternative (and less profitable) solutions, or referring business elsewhere. Buyers trust sellers who have the buyer's best interests in mind.

2. **Honor commitments.** Successful sellers earn buyers' trust by showing up and honoring their commitments consistently. Do what you promise and do it well. Make sure buyers have clear expectations for how you operate.

Intimacy

Buyers are much more likely to buy from sellers with whom they have a relationship. Strive to develop a relationship with buyers using the following methods:

1. **Create shared experiences.** Shared work experiences expose buyers to your thinking, your work style, and your work product. Also, the more time you spend with someone, the more opportunities you have to develop a solid relationship as long as you are likeable.

2. **Be a person.** Lots of sellers are told, "Don't talk about politics. Don't talk about anything personal. You can ask about the weather, but that's it. Anything else might get you in trouble." Yeah, it might, but if you don't connect with people on a personal level, you're neglecting a critical component of building trust. Don't be afraid to connect on a personal level when the opportunity presents itself.

Sellers who don't work on building trust are missing out on a powerful differentiator. Trust is a major part of the sales equation.

Five Keys to Effective Networking for Sales

Successful salespeople tend to be diligent and purposeful in networking activities. At its core, networking is about making a human connection. Utilize these five keys and you will quickly find that your network is not only expanding, but it is actively working for you to help you increase sales.

1. **Be a connector.** Do you know someone who appears to know everyone? How can you do that? It starts with meeting people. The more people you meet and learn how you can help them, the more likely you will be able to connect them to someone who is a good match. You add value for them, and they respect you for that. The other benefit is when

people you have connected sit down to talk. What is likely to be the one thing they have in common? You! It is good for business to have two people talking positively about you and your business.

2. **Listen and help.** Women are often better networkers than men, because they are typically better listeners. They know asking questions is a great way to engage with someone and draw them out. The objective of asking questions is to learn how you can help the other person—not to set them up for your sales pitch. Even though networking is not about selling, it can, of course, be a great sales tool. Why? Because networking can help you discover how you might help the other person and help them to discover how they can help you.

3. **Utilize the strength of "weak ties."** If you haven't already, you will find that most of the critical successes in your personal and professional life will come through someone who knows the person who will ultimately be responsible for that success (future client, employer, etc.). It will likely not come as a result of a direct or planned contact with that person. Those successful "friend of a friend" contacts most often happen when your "friend" is aware of who you want to meet. Thus, these seemingly weak ties can lead to significant sales over time.

4. **Follow through and follow up.** If you say you are going to do something in a meeting, that is a commitment. Do it right away. If you don't, all the hard work you put into the meeting is lost. Not only should you follow through when you are on the giving end, you should also do so on the receiving end. You will find that nothing will cause your referral sources to dry up more quickly than when you fail to contact the person to whom they recommended you reach out.

5. **Learn to tell stories.** Nothing is more engaging, or more effective, than a well-told story. That's why some of the best speeches always start with a good story. It is hard for people to remember facts and figures, but they can often recall and retell a story in amazing detail. Give an example of how you helped someone, rather than simply explaining what you do.

While networking opportunities should not be used for hard selling, networking can—if approached professionally—set the stage for sales success.

Making the Perfect 'Follow-Up Call'

Typically, it's the follow-up call that really gets the sales cycl rolling. It's here where value truly begins to manifest itself. It's here where substantive information is gathered, and it's here where the relationship begins to establish itself.

Therefore, it is absolutely vital to have superb follow-up strategies and tactics in order to make the most of the opportunity. Following are the keys to success for making a perfect follow-up call.

Tip #1: Obtain commitment for the follow-up

One of the single, biggest mistakes made by sales representatives is not establishing a specific date and time for the follow-up call at the end of the initial call.

Vague commitments from the prospects ("call me next week") or the sales rep ("I'll send the proposal and follow up in a couple of days") result in missed calls, voice mail messages, a longer sales cycle, and even missed sales opportunities. Avoiding this is a simple fix: schedule a follow-up date and time. If your prospect is not available for your initially proposed date, continue recommending other dates/times until arriving at a mutually agreeable date. If that doesn't work, get them to establish a time and date. Creating a deadline is a simple, but extremely powerful, tactic.

Tip #2: Build equity and be remembered

After every call to a first time prospect, send a thank you card. Handwrite a message on a small thank you card that simply says, "John, thank you for taking the time to speak with me today. I look forward to chatting with you further on the 16th! Kind regards…" In today's fast-paced world, a hand-written card tells the client that you took the time and effort to do something a little different. This gesture registers in the prospect's mind, creating a degree of "equity" in you. It differentiates you and gets you remembered. It also gives the prospect a reason to be there, when you make you follow-up call.

If you don't think a card will get there in time, send an e-mail with the same note. Just be aware that an e-mail does not have the same impact as a handwritten note.

Tip #3: E-mail a reminder and an agenda.

The day before your follow-up call, send an e-mail to your prospect reminding them of your appointment. In the subject line enter "Telephone appointment for August 16th and article of interest." Note that the subject line not only acts as a reminder, but is vague enough that the prospect will likely open it as there is a hint that maybe the date and time has changed.

Your e-mail should confirm the date and time of the appointment and briefly list your agenda: "Hi, John, the call should only take 10 minutes. We will review the proposal, and I will gladly answer any questions. Following we can discuss the next steps, if any. Respectfully yours, …"

The language is very similar to that used initially setting the follow up. In particular, notice the trigger phrase "…the next step." The "if any" will help reduce some of the 'stress,' pressure, or concern a first-time prospect may have. Often a first-time prospect will skip out on the follow-up call, because they are worried they will be pressured into making a commitment. This is a natural reaction. If the prospect senses an easy, informal, no-pressure type of phone call, he is more likely to follow through with the call.

Tip #4: Add value in a "P.S."

Notice in the subject line there is a reference to an article. At the end of your e-mail add a P.S. that says, "John, in the meantime, here's an article I thought you might enjoy reading…"

The article may be about your industry, the market, a product, or, better yet, something non-business related that you had discussed in your initial call. This creates tremendous value even if the client does not open it. Why? Because you took the time to do something extra, better educated them on something related to the call, and brought additional value to the subject. This helps get you remembered and gives the prospect another reason to take your follow-up call. Of course, this means you have to do some work in advance. Start looking on the web for articles of interest and value relative to your market, industry etc. Keep a file of these articles, because they can be used over and over again.

Compelling content can open doors, because it is VALUE ADDED instead of VALUE ASKING. The more you give, the more effective your later 'ask for the sale' will be.

Tip #5: Call on time

Make absolutely certain you don't start your relationship on the wrong foot. Call on time. Never be late with your follow-up call. The promptness and respect you show on a follow-up call reflects on you, your company and your products.

Tip #6: Avoid opening statement blunders

Many telephone sales representatives start off in ways counterproductive to the objective of the call. Following are a few classic follow-up opening statement blunders:

- "I was calling to follow up on the proposal."
- "I am calling to see if you had any questions.'
- "I just wanted to make sure you got my e-mail."
- "The reason for my follow-up was to see if you had come to a decision."

It is not that these opening statements are poor, but rather that they are routine, typical, and common. They do nothing to position you or differentiate you. What this really means is that you are perceived as yet another run-of-the-mill salesperson trying to make a sale. The opening statement needs to have pizzazz by creating some interest or enthusiasm and/or building some rapport with the prospect.

Tip #7: Crafting a powerful follow-up opening statement that produces results

There are four simple steps in creating a powerful opening statement. First, introduce yourself using your full name. Second, share your company name. Up to this pint in the opening it has been a simple and obvious process, but it is the third step where you differentiate yourself.

Remind the prospect of the reason for your call. This means going back to your initial cold call and informing the prospect of the "pain" or the "gain" that was discussed or hinted at in your previous call. For instance, "Cindy, this is _____ calling from ABC Company. Cindy, when we spoke last week you had two concerns: First, you indicated you were concerned about having your current on-line training program renewed automatically, before you had a chance to review it in detail, and second, you mentioned there were several modules with questionable content."

This reminds Cindy why she agreed to this call. You do this, because people are busy and have a tendency to forget. Furthermore, the urgency of last week may not seem so urgent this week.

Prospects like a clear, concise agenda. They want a reputable sales representative who is organized and doesn't waste their time. They want someone to take control and who will move the call forward. This gives them confidence.

Finally, notice how this approach repeats the theme established in the first call and in the follow-up e-mail. Remember the language used: "…determine the next steps, if applicable." It's a nice touch and reduces client resistance.

Tip # 8: Be persistent, polite, and professional, but not a pest

Be persistent in your follow up. Tenacity pays huge dividends. If someone genuinely wants you to stop contacting them, then do not make any further attempts to contact them. However, you can often avoid getting to that point by establishing good rapport and clear expectations.

If you end each conversation, regardless of how motivated the prospect is, by asking for permission to follow up with them in a certain period of time, you're always going to be operating in the realm of mutual respect and consideration when you follow up.

If they say no, then you know where you stand. If they say yes, then you have an open door and clear expectation that you will be following up. By asking for permission to follow up you lower your odds of being an annoyance and have solid footing in the future.

Summary

Having a solid follow-up strategy, with productive tactics, will separate you from the competition. It gives you a distinct advantage. Make the most of your follow-up calls, and your sales will grow.

Implement a Follow-up Schedule that Works

Don't leave follow up to chance. Most sales representatives fail to invest in setting and communicating clear expectations for what good follow-up actually looks like beyond some vague generality. *It is important to be specific and create a schedule for follow-up contacts who are most appropriate to the nature of the potential opportunity.* Create a follow-up schedule that outlines when calls and email follow-ups should be happening. Following is an illustration of how a follow-up schedule may be developed. Of course, the ideal schedule will have specific dates as well as a way to track activities for quality assurance purposes.

Time Your Sales Contact for Best Results

Not all days of the week, or times of the day are created equal. Try to time your follow up to hit the sweet spot — a time where you have the highest odds of getting a response.

Wednesday and Thursday come out on top for both email open rates and phone contact rates. In fact, according to this research:

- Thursday is 26% better for *email open rates* than the worst weekday — Monday.
- Thursday is 49% better for *phone contact rates* than the worst weekday — Tuesday.

Use Multiple Contact Formats

It's as simple as it sounds. Use more than one way to reach out. Email, snail mail, phone, text, social media — it's all on the table. The goal is to touch prospects in different ways in order to stay on their mind and stand out from the competition. Therefore, keep

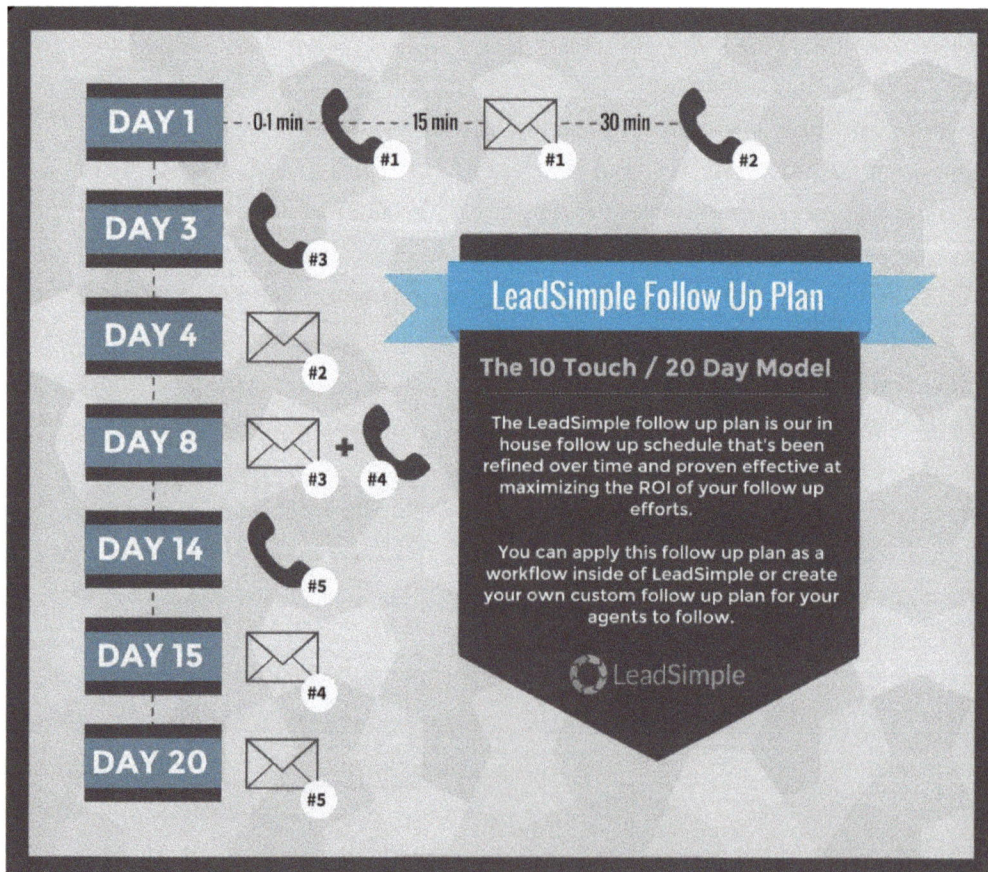

LeadSimple Follow Up Plan

The 10 Touch / 20 Day Model

The LeadSimple follow up plan is our in house follow up schedule that's been refined over time and proven effective at maximizing the ROI of your follow up efforts.

You can apply this follow up plan as a workflow inside of LeadSimple or create your own custom follow up plan for your agents to follow.

LeadSimple

in mind the best methods for reaching each of YOUR specific prospects.

Sales Success: Final Word

Sales success begins with attitude. To optimize success in sales you must understand that the sale is not about you; it's not about commission; it's not about moving an item of off the shelves; and it's not about the hefty paycheck. It is about the customer.

Getting to know your customer's interests and desires is critical to successfully closing a sale. Actively engaging your consumer to not only tap into their wants, but also anticipating their needs, are traits of a great sales person. Don't take it personal if you are unable to sell the product even after seemingly anticipating the needs of your consumer. Being successful in sales requires having a tough skin. Take rejection as an opportunity to perfect your interpersonal sales tactics and become a better sales person.

If you are engaging your customer over the phone, be upfront regarding the reason for your call, vouch for the integrity of the product or service you are selling, and draw comparisons between your prospect and satisfied customers. When leaving voicemails, emails or follow-up messages, it's important to reiterate the reason for contacting them, as well as summarizing your intent. Finish up the message with action: to make the sale and/or a follow-up engagement of some sort.

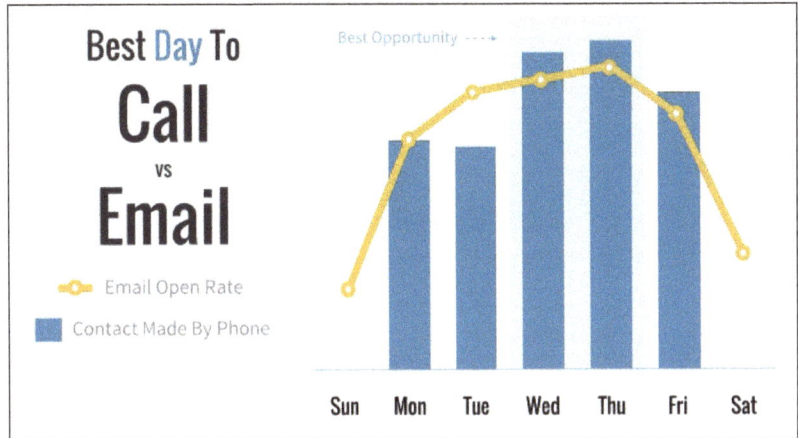

Best **Day** To
Call
vs
Email

- Email Open Rate
- Contact Made By Phone

Best Opportunity ---->

Sun Mon Tue Wed Thu Fri Sat

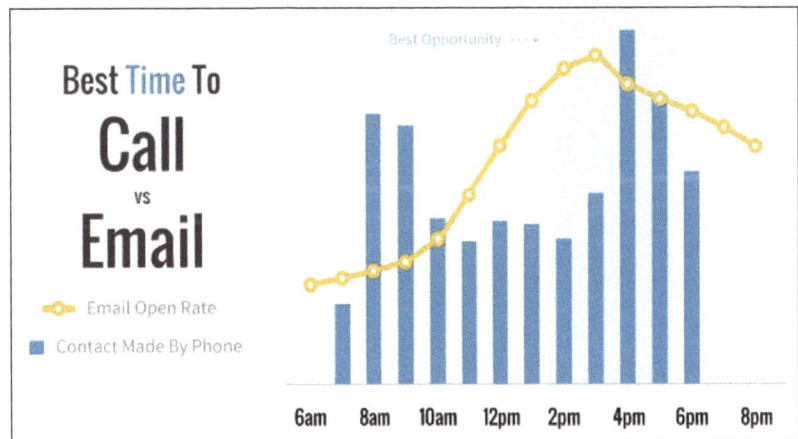

Best **Time** To
Call
vs
Email

- Email Open Rate
- Contact Made By Phone

Best Opportunity ---->

6am 8am 10am 12pm 2pm 4pm 6pm 8pm

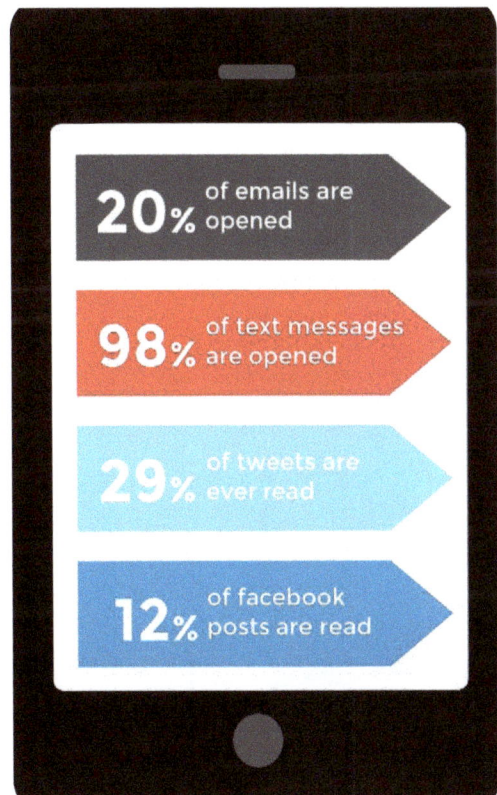

20% of emails are opened

98% of text messages are opened

29% of tweets are ever read

12% of facebook posts are read

PART II

Optimizing Success

CHAPTER 14

Email Communications: Dos & Don'ts

When to use the phone instead of email

Email serves as an excellent means exchanging information quickly, but it isn't always the most effective way to conduct business. Following are four reasons to pick up the phone instead of sending an email.

1. **You'll build key relationships.** Remember the "80-20 rule," which says the bulk of your business comes from a relatively small percentage of your customer base? Chatting with clients, customers, associates, and partners can go a long way toward understanding your colleagues' needs, preferences, operations, and temperaments — all of which are fundamental to establishing relationships.

2. **You'll cut down on confusion.** People often rely on email, because it *feels* like the quickest way to communicate. But that is not always the case. If your initial message is bound to start a long, ongoing thread of questions, then you are better off using the phone. For example, if you're kicking off a project or changing an existing system, calling people not only enables you to answer their questions immediately, but also helps you to avoid the potential misunderstandings back-and-forth email exchanges can create. If you prefer to document your conversations, use email as a follow-up tool.

3. **You will be proactive.** The spontaneous nature of phone conversations gives you the opportunity to probe deeper into the intricacies of your clients' businesses, find out your customer's needs, or dig deeper into issues your associates or suppliers may be facing. Valuable vendors are those with the unique ability to offer solutions and eliminate problems, before they escalate — email will rarely give you a true understanding of the big picture.

4. You will gain insight and negotiating power. For years, psychologists have studied how the pitch of a speaker's voice helps to reveal his or her emotional state. According to Psychology Today, men who feel intimidated raise the pitch of their voice. When people feel less dominant in a conversation, they tend to drop their pitch in an effort to assert and regain power. A speaker who is being dishonest often speaks in tones that are "deep, loud, moderately fast, unaccented, and clearly articulated." Hearing a speaker's voice pitch can improve your negotiating power, help you understand how a customer, associate, or vendor perceives your relationship, and perhaps

indicate whether there are competing factors or alarming issues—before it's too late.

Seven Reasons Why the Phone is better than Email:

1. **Phones get action.** Emails, letters, and facsimiles are often filed or placed in a stack where they are easily ignored and forgotten. With the phone, you have a much better chance of talking directly to the person and getting an immediate response. For cold calling, bill collection, and various other calls that will likely be ignored, it is beneficial to make the call from a number that doesn't reveal your information on their caller ID.

2. **A telephone conversation is much more complete than an email dialogue.** Both parties get more meaning out of a phone call than an email, because a phone call gives each the opportunity to listen to the other's tone of voice, judge the other's emotional state, and note where theother pauses and adds emphasis. By comparison, with email major talking points and important nuances are often lost. An interactive phone conversation lets you know if the person you're talking to "gets it." If not, you can use a wide range of conversational tools – pacing, follow-up questions, affirmations and repetition to communicate your message.

3. **Telephone conversations are usually less formal than emails.** A lengthy, official email may take 40 minutes to finalize, when you could complete the conversation in 10 minutes. When in doubt, jot down a few notes on the points you want to make on your call – it's still quicker than typing out and proofreading a serious email.

4. **You get instant feedback.** Does the other party agree with your point? Are they tracking with you? Are they going along with, but grudgingly? Are you going to have to adjust your pitch before contacting other prospects? The phone can give you instant, invaluable feedback that tells you how things are going and helps you get things back on track fast, when they are headed in the wrong direction.

5. **Business thrives through relationships, not through email transactions.** Email is useful for expressing a few points as concisely as possible, whereas the phone provides the opportunity to reinforce or build a relationship. An idle comment about the weather or the kids, provides an opportunity for both parties to get to know each other better and can lead to new opportunities.

6. **Getting to know someone personally gets lost in email.** To show someone how much you care, the phone is a much clearer medium for expressing emotion and conviction. Saying to someone, "I would really like to meet you" at an upcoming event is very different than typing out "I would really like to meet you."

7. **A phone call leaves no trail.** If you're contacting someone on a sensitive matter and don't really want to leave a written record, calling is better than emailing a message and asking the recipient to delete it once they have read it. We have no control over how an email may be recorded, forwarded, or even rewritten. (And yes, it's possible that someone could record or eavesdrop on a phone call, but that is harder to do, and is much more of a taboo.)

Yes, email is a convenient communication method. And the more we use it, the more we come to depend on it. However, we would be much better served to break away from the email habit. Using the telephone more frequently to accomplish your objectives will undoubtedly lead to faster, fuller, and more fruitful exchanges.

When and when not to use written communication

As discussed above, in today's world of rapid-fast communication via texts and emails, most of us would rather shoot off a written message than make a phone call. It's fast, efficient when used properly, and it provides a nice document trail for our work records. Written communication is more important than ever, yet very few people know when writing is the right – or wrong – form of communication. This section will discuss when to embrace and when to avoid written communication.

When and when not to use written communication

Complexity of the topic - Using written communication is an excellent choice for sharing information easily organized and easily understood by the recipient. This means the recipient can read the communication and get the message clearly without additional information from you or other sources. Meeting notices, answers to quick questions, or quick clarifications are all easy to complete with written communication.

There is a point, however, at which written communication becomes inefficient for one of several reasons. For instance, when the information is too complex to organize in a manner intelligible to your recipient without further assistance, it is better to make a telephone call. Any time the amount of explanation required to make the information intelligible might be cumbersome and susceptible to misinterpretation or lack of understanding, the phone call is

also a more prudent option. Lastly, when you suspect you will end up answering many follow-up emails, you are better off having a face-to-face meeting or a phone call.

Amount of 'discussion' required - If the topic is complex or involved enough to require a long exchange of discussion-type emails, you are risking someone misunderstanding the point. Furthermore, this typically evolves into a major time drain - preparing, reading, responding to, and filing emails.

Additionally, you can't be assured everyone who received the email has actually had the chance to participate in the discussion, unless you are able to track the receipt of others' emails or require everyone to respond one way or the other. Therefore, decision-making, long, involved explanations or conversations, or controversial subjects are not usually good topics for written communication.

Shades of meaning - Non-verbal communication is the most important form of communication in getting your message delivered. When you are writing, you are left to the small portion of communication possible through words alone in getting your message through to your reader. So the more intense the emotions around a topic or the more important the message the less likely writing will be a successful form of communicating.

For example, it can be difficult to convey tone of voice, humor, sarcasm, or other shades of meaning in writing alone. We risk offending someone or causing confusion through written communication. Conveying highly emotional or important information is better accomplished via telephone or in person.

Formal communication - Written communication is best used as a formality between two parties. Formal communication - contract terms, sales agreements, account information, schedules, meeting notifications, legal or administrative information - is best conveyed in written form.

Tension - The moment you sense tension in an email you need to call the other party. The best approach is to pick up the phone and speak with the other party in a transparent and caring way that leads to an agreed upon solution.

Confusion – It is beneficial for both parties to call the other person if you don't understand what is being communicated to you in an email. Furthermore, it is time to make a phone call when you sense the other party doesn't understand what you are trying to communicate in an email. The moment the dial on the confusion-o-meter starts to go up, pick up the phone.

Celebration – Make a call to the other party, when there is something to celebrate. This may seem a little goofy, but sharing good news produces positive feelings and helps build relationships. Sharing good news is a great opportunity to strengthen customer loyalty, enrich working relationships, and win favor with superiors.

Decisions - Key decision-making is often accompanied by feelings of hesitancy and fear. Making a call helps the other person move past negative, immobilizing feelings and onto making a decision. Therefore, ditch the email and make a phone call.

Calling a client on the brink of making a key decision gives that person an opportunity to share his or her concerns and gives you the opportunity to and lead him towards a favorable decision. It is mission critical for you to recognize when your clients need to make key decisions, and at these points of engagement you must be willing to pick up the phone.

Critical feedback – It is prudent to meet in person or make a phone call when giving critical feedback, especially those of serious, performance-based, or personal nature.

Delivering difficult or sensitive news – It is best to schedule a meeting or make a call whenturning someone down for a raise or promotion, discussing concerns about attendance, or ending someone's pet project.

Questionable issues – If you're dreading the conversation, or it feels uncomfortable to you, then email is probably not be the best approach. Those feelings should serve as a sign that the issue is sufficiently delicate, emotionally charged, or ripe for misinterpretation, and meeting in person or speaking on the telephone may be more productive.

In Conclusion - You can still write emails to back up a phone call. However, a phone call instead of an email, in the situations mentioned above, will produce the best results and often save a lot of time in the long-term.

When to Send an Email

As previously stated, there are many situations when it is best to schedule a meeting or make a telephone call. However, email is a hugely valuable communication tool. Email is inexpensive, only requiring an Internet connection generally already present in the business and readily available for most people. Although a printout of emails is possible, emails often stay as soft copies, because archiving and retrieving email communications is easy to do. Most everyone from the CEO to the janitorial staff and temporary employees of a business can send and receive email communications.

Internal emails can function as an effective communication for sharing basic information, such as new cafeteria prices, paper use guidelines or security precautions, for example. Sending simple messages to an entire workforce with just the click of a mouse is fast, easy, convenient and can save the company money. If saved, the email can function as proof of a message sent or received and is easily accessible to remind the recipient of pertinent information. Many businesses use email as part of their marketing efforts to share information with prospects, customers, vendors.

Even so, emails should only be used as a communication tool in certain situations. Following are the best practices guidelines as to when to use email.

1. **Communicating schedule and straight-forward information** – Meeting notices, answers to quick questions, or quick clarifications are all easy to complete with written communication.

2. **Non-Urgent Communication** – Email is great for non-urgent communication. Things that don't require an immediate response and others can deal with on their own schedule.

3. **Follow-up** – Email can provide a powerful documentation trail. Unlike text messages or phone calls, email provides a very concrete audit trail of past communication. It is hard to deny past actions and messages when there is a clear history.

4. **Praise** – We all know that it can be dangerous to send a negative email. They often get misinterpreted for tone or meaning. However, emailing *praise* is very powerful. Almost everyone enjoys opening an email to find an encouraging or motivational note.

5. **Time-shifting** – Email is one of the best mediums for "time-shifting" of communication. It allows people on different schedules, or even time zones, to communicate at their leisure.

6. **Filtering** – Phone calls and text messages are difficult to screen. However, this is an area where email excels. You can set up filters to prevent unwanted email from even reaching you.

7. **One-to-Many Communications** – Email provides a very efficient and productive way to communicate non-urgent items to large numbers of people.

8. **Sending Documents/Pictures** – With the advance of scanners and other technology, there is very little reason to send physical documents. Email is very powerful for sending documents, pictures, and more.

9. **Mobility** – This can be a double-edged sword, but if used correctly, email can actually be a liberator. With mobile access to email, we are no longer tied to a physical desk or location. I can work from the local cafe or the beach.

10. **Written Record** – Email is useful when you want a written record of what was said – to refer back to later or provide documentation of what was communicated.

11. **Instructions** – Email is useful when it is necessary to supply complicated information for which it is helpful to have details in writing, such as a new procedure for database entries or login instructions for your website.

Email Etiquette

The average US employee spends about a quarter of the work week combing through the hundreds of emails we all send and receive every day.

But despite this fact, a large percentage of professionals *still* don›t know how to use email appropriately. In fact, because of the sheer volume of messages we›re reading and writing each day, we may be more prone to making embarrassing errors—and those mistakes can have serious professional consequences.

In the age of the Internet, you might find yourself clicking "reply," typing up a quick response, and hitting "send" without giving so much as a thought to what you've just written. But such e-mail behavior has the potential to sabotage your reputation, both personally and professionally.

Following is a list of modern email etiquette of the most essential rules we all need to know. Abiding by these rules will produce many benefits, such as more time in the day, more respect as a true professional, and greater opportunities as those communicating with you will appreciate your communication efficiency and effectiveness.

1. **Include a clear, direct subject line**—People often decide whether to open an email based on the subject line. Choose one that lets readers know you are addressing their concerns or business issues. Most importantly, your subject line must match the message. Furthermore, never open an old email, hit "reply," and send a message that has nothing to do with the previous one. Do not hesitate to change the subject line as soon as the thread or content of the e-mail chain changes.

     With inboxes being clogged by hundreds of emails a day, it's crucial that your subject line gets to the point. It should be reasonably simple and descriptive of the contents of the email.

2. **Use a professional email address**—If you work for a company, you should use your company email address. But if you use a personal email account—whether you are self-employed or just like using it occasionally for work-related correspondences —-you should be careful when choosing that address.

     You should always have an email address that conveys your name so the recipient knows exactly who is sending the email. Never use email addresses that are not professional and appropriate for use in the workplace, such as "babygirl@...," " meathead@...," or "hahaha@....."

3. **Think twice before hitting "Reply All"**—Do not hit "Reply All," unless every member on the email chain needs to know. You want to make sure you are not sending everyone on a list your answer when only a few need to know.

     No one wants to read emails from 20 people that have nothing to do with them. Ignoring emails can be difficult with many people getting message notifications on their smart phones or distracting pop-up messages on their computer screens. Refrain from hitting "Reply All," unless you are confident everyone on the list needs to receive the email.

     Before you click "Reply All" or put names on the "Cc" or "Bcc" lines, ask yourself if all the recipients truly need the information in your message. To ensure the right people receive your meesage, beware of the "Reply All," "Cc," and "Bcc" buttons.

4. **Include a signature block**—You never want someone to have to look up your contact information to get in touch with you. Your email signature block is a great way to let people know more about you, especially when your e-mail address does not include your full name or company. The ideal signature block should state your full name, title, company name, and your contact information, including a phone number. You also can add a little publicity for yourself, but don't go overboard with sayings or artwork.

5. **Know your audience and use professional salutations**—Your email greeting and sign-off should be consistent with the level of respect and

formality of the person with whom you're communicating. Always keep your recipient in mind - if they tend to be very polite and formal, write in that language. The same goes for a receiver who tends to be more informal and relaxed.

However, don't use laid-back, colloquial expressions such as, "Hey you guys," "Yo," or "Hi folks." "The relaxed nature of our writings should not affect the salutation in an email."*Hey*" is a very informal salutation and should not be used for professional communications. Use *Hi, Dear,* or *Hello* instead.

Furthermore, avoid shortening anyone's name. For instance, communicate with "Hi Michael," unless you're certain he prefers to be called "Mike."

6. **Use exclamation points sparingly**—If you choose to use an exclamation point, use only one to convey excitement. Some people have a tendency to put too many exclamation points at the end of their sentences. The result can appear to others as being too emotional, immature, or unprofessional. Exclamation points should be used sparingly in writing.

7. **Be cautious with humor**—Humor can easily get lost in translation without the right tone or facial expressions. Something perceived as funny when spoken may come across very differently when written. Furthermore, something you think is funny may not be funny to someone else and may be viewed as a waste of valuable time to many who feel overloaded. In a professional exchange, it's better to leave humor out of emails, unless you know the recipient very well.

8. **Know that people from different cultures speak and write differently**—Miscommunication can easily occur because of cultural differences, especially in writing, when we can't see one another's body language. A good rule to keep in mind is that high-context cultures—-such as Japanese, Arabic, and Chinese—-want to get

to know you before doing business with you. Therefore, it may be common for business associates from these countries to be more personal in their writings. On the other hand, people from low-context cultures—such as Germany, America, or Scandinavian—prefer to get to the point very quickly.

9. **Reply to your emails; even if the email wasn't intended for you**—It is difficult to reply to every email message ever sent to you, but you should try to do so. This includes emails accidentally sent to you, especially if the sender is expecting a reply. A reply isn't necessary but serves as good email etiquette, especially if this person works in the same company or industry as you.

Here's an example reply: "I know you're very busy, but I don't think you meant to send this email to me, and I wanted to let you know, so you can send it to the correct person."

Replies do not have to be long and/or time consuming. Actually, the shorter you can keep it, the better.

10. **Proofread every message you write before hitting the 'send' button**—Your mistakes won't go unnoticed by the recipients of your email. Don't rely on spell-checkers. Read and re-read your email a few times, preferably aloud, before sending it. Your emails directly reflect who you are to others on both a personal and professional level.

Every email you send adds to or detracts from your reputation. If your email is scattered, disorganized, and filled with mistakes, the recipient will be inclined to think of you as a scattered, careless, and disorganized business person, and other people's perception of you is critical to your success.

11. **Add the email address last**—Even when you are replying to a message, it's a good precaution to delete the recipient's address and insert it only when you are sure the message is ready to be sent. You never want to send an email accidentally,

before you have finished writing and proofing the message.

12. **Double check that you've selected the correct recipient**—Pay careful attention when typing a name from your address book on the email's "To" line. It is far too easy to select the wrong name which can be embarrassing to you and an inconvenience to the person who receives the email by mistake.

13. **Keep your fonts classic**—For professional correspondence, keep your fonts, colors, and sizes classic. There are approximately 40 classic fonts. However, according to *Business Insider* the two best professional fonts for email communications are Georgia and Verdana. Both of these fonts have even and appropriate letter spacing, as well as a letter stroke style that makes the text very easy for the eye to read.

   Generally, it is best to use 11- or 12- point type. Although, the 12-point type text is appreciated more by those recipients with less than perfect eye sight. Type should be black in color.

14. **Keep tabs on your tone**—Just as jokes get lost in translation, tone is easy to misconstrue without the context you would get from vocal cues and facial expressions. Accordingly, it is easy to come off as more abrupt than you may intend. For

instance, your straightforward may be interpreted as angry and curt.

To avoid misunderstandings read your messages out loud before sending them. If anything in your message sounds harsh or questionable, it will probably be perceived as harsh to the recipient.

Furthermore, avoid using unequivocally negative words such as "failure," "wrong," or "neglected." Seek a better way to communicate these impressions. Also, use the word '"lways" with much caution, but use the words "please" and "thank you" abundantly.

15. **Nothing is confidential—so write accordingly.** Never forget that *every* electronic message leaves a trail. A basic guideline is to assume others will see what you write, so never write anything you wouldn't want everyone to see. Most importantly, never write anything that would be ruinous to you or hurtful to others. We can be assured that NSA archives every email sent, which in turn becomes part of our record. Also, email is too easily forwarded or saved by others. It is much better to play it safe than to have something come back to negatively impact you or your reputation later.

We have all heard the stories about a "private" email that ended up being passed around to the entire company, and in some cases, all over the Internet. One of the most important things to consider when it comes to email etiquette is whether the matter you are discussing is a public one or something that should be talked about behind closed doors. Ask yourself if the topic being discussed is something you would write on company letterhead or post on a bulletin board for all to see before clicking the "send" button.

16. **Briefly introduce yourself**—If you are uncertain whether the recipient recognizes your email address or name, include a simple reminder in your message. Do not assume the person receiving your email knows who you are or remembers

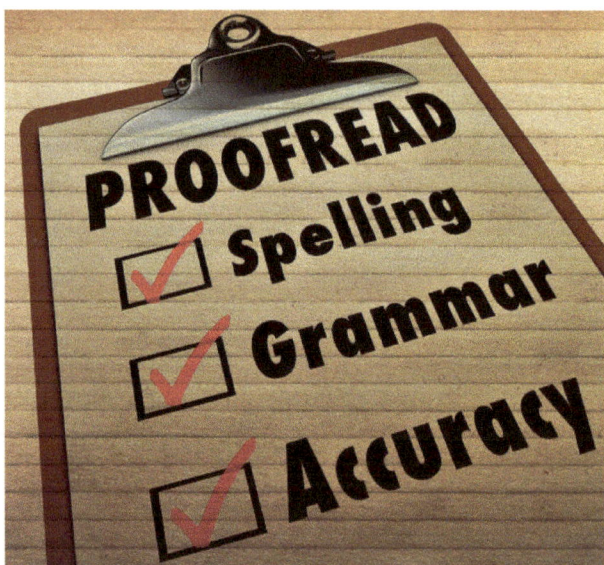

meeting you. However, an extensive biography is not necessary and should not be included.

17. **Don't email angry** –Emails conveying bad news, firing a client or vendor, expressing anger, reprimanding someone, disparaging other people, or saying anything less than kind about another person are all major no-no's. Because email can seem so informal, many people fall into this trap. Keep in mind that email correspondence lasts forever.

18. **Be careful with confidential information.** Refrain from discussing confidential information in emails such as someone's tax information or the particulars of a highlysensitive business deal. Should the email get into the wrong person's hands, you could face serious—even legal—repercussions.

19. **Be reasonable about responding to emails.** Unless you work in some type of emergency capacity, it's not necessary to be available the instant an email arrives. Checking your email regularly during the day can be an effective way to keep your inbox at manageable levels. However, the constant interruption and distraction that comes from multitasking in this way can dramatically lower your productivity and disrupt your ability to enter a state of flow when working on high value projects. One strategy you can use is to check email only at set points during the day. Typically, depending on the nature of the email and the sender, responding within 24 to 48 hours is acceptable.

20. **Reduce reply emails.** Place "No Reply Necessary" at the top of the email, when you don't anticipate a response.

21. **Avoid using shortcuts to real words, emoticons, jargon, or slang.** Words from grown, business people using shortcuts such as "4 u" (instead of "for you"), "Gr8" (for great) in business-related email is not acceptable. If you wouldn't put a smiley face or emoticon on your business correspondence, you shouldn't put it in an email. Any

of the above has the potential to make you look less than professional.

22. **Keep it clean.** Many people are annoyed and/or view the sender negatively, when they receive a messy email. For example, an email chain that includes excessive carets (>>>), or pages and pages of email addresses that weren't protected by using Bcc are considered sloppy. You can get rid of carets by selecting the text, Ctrl+F to use the Find and Replace command to find a caret and replace all of them with nothing. Also, take care to delete irrelevant email addresses in the email you are forwarding. In summary, clean up emails before you send them.

23. **Don't get mistaken for Spam.** Avoid subject lines that are in all caps, all lower case, and those that include URLs and exclamation points. Such emails are easily mistaken as spam by recipients.

24. **Alert the recipient in advance of sending large attachments.** Sending unannounced large attachments can clog the receiver's inbox and cause other important emails to bounce. If you are sending something that is over 500KB, you should ask your intended recipitent, "Would you mind if I sent you an attachment? When would be the best time for you?" Another approach is to send an email to the recipient alerting them that you are forwarding a large file(s) to them via an online file sharing service such as Dropbox or We Transfer.

25. **No more than two attachments, and provide a logical name.** Unless it's been specifically requested, refrain from sending a message with numerous attachments. Also, give the attached file(s) a logical file name so the recipient knows at a glance what to expect within each attachment.

26. **Use the telephone instead of email when appropriate.** When a topic has lots of parameters to be explained or negotiated and will generate too many questions and confusion, don't handle it via email. Email should not be used for last minute

cancellations of meetings, lunches, interviews, or for devastating news. If you have an employee or a friend to whom you need to deliver bad news, a phone call is more appropriate. Email is more practical for sharing news and information with a large group.

27. **Evaluate the importance of your email.** Don't overuse the "high priority" option. As with crying wolf, overusing this feature, will cause few people to take it seriously, and many will view you as an emotionally anxious person. A better solution is to use descriptive subject lines explaining the specific content of your message.

28. **Maintain privacy.** If you're sending a message to a group of people, and you need to protect the privacy of your list, you should always use "Bcc." Likewise, avoid giving out email addresses to a third party (such as an Evite, newsletter, etc). Make sure any third parties to whom you give addresses are reputable and reliable entities. Be especially cautious regarding third parties who offer free services.

29. **Keep it short and get to the point.** The long email is a thing of the past. Write concisely to conform with professional best practices and to be respectable of the recipient's time. Make sure what you're sending isn't a burden to read. The use of bullet points is an excellent and acceptable way to shorten an email. The person reading your email should not have to read through several paragraphs to determine what you are communicating or asking. State the purpose of the email within the first two sentences. Be clear, concise, and up-front.

30. **Only use an auto-responder when absolutely necessary.** An automatic response that says, "Thank you for your e-mail message. I will respond to you as soon as I can" is useless. Instead prepare an auto-response that specifically communicates what can be done in your absence and/or when you will reply. Auto-responses should be used sparingly as spammers will capture, use, and share your email address.

31. **Train your staff.** Business owners and managers should educate and train their staff in email communications. Furthermore, email rules and policies should be established for everyone within the company to follow.

> **Be sure to read the "Time Management" chapter within this book, as it also includes many helpful email tips.**

In Summary

Email is a good way to alert people of key news and direct them to detailed information elsewhere. If you need to cover several topics, summarize the important points in the first paragraph and provide highlights with subheadings and brief introductions that link to attachments or intranet pages for more information.

Business emails should be concise and to the point. Use plain text and common fonts with a simple signature line. Fancy graphics, fonts, and backgrounds can take up unnecessary storage space in the recipient's inbox and may load slowly, or not at all. Stick to one topic in a business email and write only information appropriate for anyone to read, as email forwarding makes it possible for originally unintended parties to receive the email. Proper grammar and spelling is very important in business emails, because it reflects on you, your abilities, and your company. Attachments should be in a format any recipient can easily access/download.

Whether sending an internal or external email, the subject line should accurately represent the content of the email. Use proper punctuation and capitalization and use bold, italics and white space to make important information stand out from the rest of the content. Use a salutation to begin the email and only send to individuals who specifically need or request to receive information from your business. Business emails should be brief, positive and professional.

Using email in business communication is certainly less personal than face-to-face communication. It can hinder social interaction and lead to less relationship building in the workplace. Misunderstandings are commonplace in written communication simply because the recipient cannot read the writer's tone and expression. Consider that even though email can be quite informal, it is also a permanent record and should be treated carefully.

Email has been revolutionary in the world of business communication, because information is quickly passed along with instantaneous efficiency and effectiveness. Employees are able to access information from a computer, phone or PDA nearly anywhere, and so are the company's current and potential customers. The use of email within a business can greatly increase productivity for employees and can be a quick way to increase sales as well. More companies are using email communications as part of the marketing mix to communicate and interact with its target markets.

Email is an excellent tool when used effectively. Use the information provided within this chapter, and the Time Management chapter, to utilize email more productively.

CHAPTER 15

Networking for Increased Value, Rewards, and Opportunities

What is Networking?

Networking is an activity in which business people and entrepreneurs meet to form business relationships and to recognize, create, or act upon business opportunities, share information and seek potential partners for ventures. A business network allows business people to connect with other professionals and further each other's interests by forming mutually beneficial business relationships.

Many business people contend business networking is a more cost-effective method of generating new business than advertising or public relations efforts, because business networking is a low-cost activity that involves more personal commitment than company money.

Business networking can be conducted at the local community level, the regional level or the national or international level. The Internet and teleconferencing services make it possible for business people from similar industries or sectors to connect, even if they live in different regions or countries. Business networking websites have grown with the Internet.

Benefits of Networking

Business Networking is a valuable way to expand your knowledge, learn from the successes of others, attain new clients and market your business. Following are detailed benefits of networking:

1. Increased Business

This is probably the most obvious benefit and the main reason most professionals engage in networking activities and join networking groups.

The fact is referrals you receive through networking are normally high quality and typically even pre-qualified for you. These referrals can often be easily converted to clients. Networking referrals usually result in much higher quality leads than other forms of marketing; therefore, the increase in business from networking is a tremendous benefit.

2. Opportunities

There are always lots of opportunities available through networking. In fact, this is where the benefits

of business networking really pay off in the form of opportunities such as joint ventures, client leads, partnerships, speaking and writing opportunities, business or asset sales, professional development, friendships, etc. The opportunities that arise from networking are almost endless.

You just want to be sure you are selective in taking advantage of the right opportunities and avoid the trap of falling into every opportunity that comes your way. The opportunities in which you get involved should align with your goals, interests, and vision. If not, you will find they become not only a distraction but a drain on your energy and time.

3. Connections

As the saying goes, "it's not WHAT you know, but WHO you know." This is so true in business. To truly optimize success in business, you need to have a great source of relevant connections in your network on whom you can call when you need them.

Networking provides you with this great source of connections. It also opens the door to highly influential people to whom you may not otherwise have access.

It's not just about with whom you are directly networking– that person will already have a network you can now tap into as well. Therefore, don't hesitate to ask the right questions to find out if the person with whom you are networking knows who you want to know and can help you make the connection.

4. Advice

Networking is a great way to tap into advice and expertise you wouldn't otherwise be able to obtain elsewhere or without considerable cost. Communicating with other professionals gives you the opportunity to receive advice from them on all sorts of matters related to your business or even your personal life.

Without counsel plans fail, but with many advisers they succeed. Just make sure you are getting sound advice from a qualified person—someone who actually knows what what you need to know and is not just giving you their opinion on something with little or no expertise on the matter.

5. Improved Reputation

Being more visible and getting noticed is another benefit of networking. Regularly attending business and social events will help get your face known. You can then build your reputation as a knowledgeable, reliable and supportive person by offering useful information or tips to people who need it. Through these contacts you will get more leads and referrals, as you will be the one who comes to mind when they need what you offer.

6. Positive Influence

Positive, uplifting people help you grow and thrive as a person and as a professional. Networking is a great way to bring these positive influences into your life. Walk with the wise and become wise.

7. Increased Confidence

Confidence is increased by through networking as it pushes us to step out of our routines and comfort zones. For example, as you get in the habit of talking to people you don't know, you gain more and more confidence.

Networking results in professional and personal growth as we learn how to make conversations and lasting connections with people outside our normal circles. The more you do it, the more confident you will become. With continued practice and increased confidence, talking to people you don't know becomes easier and easier.

8. Satisfaction from Helping Others

Networking is a fantastic way to help others. Through networking you will meet people who have business problems, and you will receive great satisfaction from helping others find solutions to their problems or helping them better manage those difficult issues.

9. Friendship

Many friendships form as a result of networking with others. As you meet interesting people and help each other, a friendship naturally forms.

10. Job Leads

Networking can produce job leads and put you in touch with people directly responsible for hiring for their company. Sometimes networking is conducted for this purpose, other times it is a byproduct (i.e. someone you meet tells you about an opening and encourages you to apply). Either way, it is a positive reward of networking efforts.

Methods and Keys to Effective Networking

Effective networking is the linking together of individuals who, through trust and relationship building, help one another in various ways.

Prioritize you networking efforts. Clubs, workshops, associations, and events are based on a particular purpose. Focus on those that most closely align with your goals. The most important thing to keep in mind is your objectives for networking. What are you trying to gain? What do you have to offer? When you have clear answers to these questions, you will better be able to focus on making quality connections and staying on track with your goals.

Make a positive impression. People want to work with positive people. Having an upbeat attitude and response will increase your chances of creating lasting impressions. Just remember to be genuine, don't be overly cheery and gung-ho. People will connect with you more when you're positive but still down to earth.

Be genuine and authentic. The most effective networking consists of building trust and relationships which often coincides with demonstrating how you can help others. People easily discern and are quickly turned off when the other person is only concerned about himself/herself.

Visit as many groups as possible that align with your goals and interest. Notice the tone and attitude of the group. Are people supportive of one another? Does the leadership appear competent? Are there ample opportunities to be involved, interact with others, and develop relationships?

Focus on quality, not quantity. Instead of focusing on building relationships with as many people as possible, focus on quality connections. Is there a great chance for a mutual benefit? If not, move on. If there is, try to find a way to help the other person succeed in reaching his/her goals.

Seek out opportunities to serve and volunteer for positions within the organizations. This is a great way to stay visible and give back to groups that have helped you.

Be interested in others. The best way to make friends and network is to become interested in others. Seek first to understand, then to be understood. People will be more attracted to you, if they know you care about them and that you are seeking a relationship based on more than just what is good for you.

Ask open-ended questions in networking conversations. This means questions that ask who, what, where, when, and how as opposed to those that can be answered with a simple yes or no. This form of questioning opens up the discussion and shows listeners you are interested in them.

Actively listen to others. Most of the time when we are listening to another person, we tend to think about what we are going to say next. Instead of just thinking about your response, quiet your thoughts and really listen to what the other person is saying.

Uncover their needs. Try to determine how you can benefit the other person. Ask them questions about their goals and aspirations.

Offer help. Once you have uncovered a perceived need, offer your assistance. Even if they decline, they will appreciate your willingness to help and will be more likely to offer you assistance in return.

Become known as a powerful resource for others. When you are known as a strong resource, people remember to turn to you for suggestions, ideas, names of other people, etc. This keeps you visible to them.

Have a clear understanding of you. Know what you do; why you do it; for whom you do it; and what is special or different about your way of doing it. In order to make helpful contacts and obtain valuable referrals, you must first have a clear understanding of your objectives and what you do in order to be purposeful in your networking and able to easily articulate what you are about to others.

Be able to communicate your needs and how others may help you. Being ready to express your mission, and specifically asking for help, when the opportunity presents itself will provide the best results.

Follow through quickly and efficiently on referrals. When people give you referrals, your actions are a reflection on them. Respect, honor, and show sincere appreciation to those contacts, and your referrals will continue to grow.

Call those you meet who may benefit from what you do. Express that you enjoyed meeting them and suggest getting together to share ideas.

Best Places to Network

The best places to network are where you can easily meet people. Places and events that share a common denominator make meeting people easier. The best places, however, are those that align with your goals. Following are excellent networking ideas for meeting people.

Industry Events—Every industry has events mainly attended by people involved in the industry. These allow you to explore new innovations in your field and interact with peers who may be years ahead of you. These events allow you the opportunity to connect with new strategic partners and joint ventures. Attending events where you are surrounded by those who are in a similar business allows you to discuss changes in the industry and exchange information on new insights that may improve your level of service and profitability.

Meetups—Browse Meetup.com and find groups that interest you then start attending those meetups. Stick with the ones that best align with your objectives. Meet people, distribute your business card, and follow up with those you meet.

Learning Events—Attending events where your primary goal is to learn something new is a great investment in your business. Gaining a new perspective and discovering new strategies will speed your growth when you attend events outside your industry. Attending these events can be thought of as a strategic advantage. You will learn what other professionals are doing and can implement those innovations that will work best for your situation.

scrapbooking, drawing, painting, sculpturing, and any other creative form can be done in a group setting with other people.

Hobby Clubs—If you have a hobby, finding a local club of people who do the same thing is a fun and productive way to network. You can meet people, share tricks and tips, and build relationships.

Breakfast Shops—Some breakfast and coffee shops are hotbeds for meeting people and developing relationships. Check out breakfast and coffee shops near you to see if they have a regular crowd that congregates.

Neighborhood Groups—Homeowners' associations and neighborhood watches can be good ways to meet people. A productive neighborhood project, like organizing a block party or setting up shared child or pet care between yourself and your neighbors, can also turn into a wonderful networking opportunity.

Conferences—A conference is the perfect place to get to know people and swap information. If you are reluctant to approach people, giving a presentation or speaking on a panel are good ways to have people approach *you*. The same is true if you serve in a role at the conference such as working at the information booth or helping with the check-in. In these roles people approach you, and you will be more visible to everyone.

Ethnic or Gender-Based Clubs—Sharing a common heritage makes it easier to find both common ground and trust. Clubs based on heritage and gender issues abound. Although it is easier for women to find gender-based clubs and organizations than for men to do so, there are plenty of official and *de facto* male organizations out there, as well. With this form of networking you can make friends based on who you are and from where you came.

Religious Events—A shared belief system like religion is an incredibly powerful, unifying force. You become part of something special - analogous to a tribe or club. This exclusionary aspect solidifies the bond between members of the belief-based organization. Trust is formed among members which results in people opening up more and having a desire to help each other.

Ideal Client Events—These are the events where the primary attendee is your ideal client. Ideal client events expand your reach beyond your current network. These events are full of prospects and are ideal for sales and business development. In these types of events you meet people who need what you have to offer.

Exercise—Whether you do team sports or individual sports done in a group, like a running club, exercise is an excellent way to get to know people outside of work. Playing together—or, in the case of endurance sports like road biking, suffering together—builds camaraderie and respect among members of the group. Health clubs and gyms are also great places to meet people and build relationships while getting in better physical condition.

Arts—This category could include performing arts, theater, improvisation comedy, knitting clubs,

Retreats—Retreats allow you to get together with groups of people who have the same interest or goal. These groups can be small (only a few people) up to a large group (several hundred participants.) You are typically with each other from one day to several weeks. With that kind of intimate setting, it is hard not to find out who people are, what they do, and how you might be able to help one another.

Social Media—Social media can be beneficial or futile. On the one hand, you can waste a tremendous amount of time reading and posting comments. On the other hand, if you link up with the right people and engage with the right mix of followers, social media can be a gold mine. When done right you can find jobs, business partners, business opportunities, services you need, and much more. The opportunities are limitless. You just have to approach it strategically and focus on continually engaging in areas that produce results.

Classes—If you are learning something along with everyone else, meeting people and developing relationships comes naturally. You can help each other understand a concept, share books or other relevant resources, participate in a study group, or discuss the material being presented. The class teacher may also be a good person with whom to network.

Associations – There are hundreds, if not thousands, of associations. When you join an association you share with others the common bond of membership. The trick is to attend the association's events and get involved at a local level.

Volunteer Work – Volunteering makes it easy to bond with other volunteers over a common cause, even if your background, religion and political views are not the same. Volunteering gives people a good feeling, which makes all parties more approachable.

Toastmasters—Facing one of humanity's deepest fears, public speaking, with twenty or so supportive new friends, who are going through or have gone through the same thing, is an excellent way to network. Everyone is motivated and eager to help one another. There are lots of opportunities to share and get to know people. Toastmasters provides lots of networking opportunities and is a great way to develop positive relationships.

Trade Shows—Trade shows are an easy and obvious place to meet people relevant to your interest or objective. If you want people to come to you, work a booth, otherwise spend time exploring other peoples' stations and networking along the way. Afterwards, follow up with people you met at the trade show though phone calls, emails, or social media.

Pets – Becoming involved in dog training classes, agility courses, therapy pet associations, or even at a dog park is a great way to network.

Kids' Events and Parent Circles – If you have children, then you have an abundance of opportunities to meet other people. The key is to get out of your comfort zone by breaking away from the people with whom you usually talk, introduce yourself to others, and sincerely try to get to know them. Everyone has a story and knows other people. It is amazing how many opportunities these casual encounters can produce.

Alumni Events—High school reunions provide good networking opportunities. College alumni events, breakfasts, receptions, alumni sports games, alumni interest groups, and alumni databases are good ways to connect under a common umbrella. Giving talks at events, or offering to help with reunion planning duties, will provide even more networking opportunities.

Speeches and Talks—Attending a speech given by your favorite politician, author, astronomer, or any other person you find to be interesting can be an excellent networking opportunity. Everyone attending typically shares the same interest , and you will have the opportunity to chat with people sitting next to you before and after the event. You also may be able to meet the speaker after the event and additional people while waiting for autographs.

Chambers of Commerce—Most Chamber of Commerce organizations hosts classes and events, as well as networking events. These are the kind of place where you can meet many people in one setting and collect a lot of cards. Teaching a class, giving a talk, or getting involved with a Chamber effort is also a great way to make connections

Leads Groups and Networking Events—These groups can provide you with a treasure chest of business. However, only by following up with your leads do networking events pay off.

Blogging—Blogging can be a good way to get to know people with common interests. You can find a community of like-minded bloggers on the web. Acquaint yourself with them by reading their posts and becoming a regular, proactive commenter. Facebook, Twitter, LinkedIn, and other sources of social media are also great ways to reach people.

Music—Music, whether you play or listen, has a wide array of networking possibilities. From playing in a band to attending certain concerts, from music festivals to the symphony, opportunity abounds for finding people with similar musical interests.

Online Forums—Online forums are places to ask questions and discuss topics of interest. Becoming a regular or a moderator on a forummakes you recognizable to the other people who frequent the site. This opens the door to allow you to get to know one another on a more personal level. Since many people prefer to stay anonymous on forums, this approach requires more persistence than some of the other options listed.

The 'In the Know' People—Some people have huge networks. Whenever you mention you need something, they automatically link you with someone in their network who can help you out. Get to know these people and stay in touch with them.

CHAPTER 16

Interpersonal & Human Relationship Skills

Interpersonal skills are the life skills we use every day to communicate and interact with other people - both individually and in groups. People who have worked on developing strong interpersonal skills are usually more successful in both their professional and personal lives than those who have ignored these skills. To succeed in business you need good interpersonal skills to understand how to deal with other people.

Some examples of interpersonal skills include:

- Communication skills: involves both speaking and listening effectively.
- Assertive skills: requires self expression without offending or violating the rights of others.

- Conflict resolution skills: dictates effectively resolving the differences that impede the formation of relationships.
- Anger management skills: involves identifying and expressing anger appropriately in order to solve problems, handle emergencies and achieve goals.

Tips to Improving Your Interpersonal and Human Relations Skills

Interpersonal skills are invaluable in all areas of life; however, in the business world they are essential for success. How your coworkers see you has a big impact on your career, as well as your day-to-day work life. Let's explore ten ways to improve your interpersonal skills.

You may be the most brilliant person at your company, but if you can't get along with your colleagues, you won't get far. Fortunately, there are several things you can do to strengthen your social skills, be a better team player, and be well-liked by others. These 10 tips will not only help you make better connections at work, they'll also improve how others perceive you.

Interpersonal Skills	
Communication	Influence
Leadership	Motivation
Negotiation	Problem solving

Tip #1: Put on a happy face

People who are the life of the party usually have one thing in common: They're happy. If you smile often and have an upbeat attitude, your coworkers will be drawn to you. On the occasion when you may be having a bad day, don't try to pull others down with you. People are likely to pass you by in favor of those with a more cheerful outlook.

Tip #2: Show that you care

When it comes to praise, don't hold back the applause. When someone does something you appreciate — no matter how small — praise and/or thank them for it. Identify at least one attribute you value in the other person and tell them about it. Give others a warm welcome when they call or visit you. Showing others how much you care about them encourages them to do the same in return.

Tip #3: Be considerate and mindful of others

Take note of what's happening with others. Recognize the happy events in their lives — from a birthday to a kid's kindergarten graduation — and be sure to show genuine compassion when they face any personal tragedy. Look people in the eye when you speak to them and refer to them by their first names. Show others you value their viewpoint by asking for their input and seeking their opinion on issues.

Tip #4: Be an active listener

Being an active listener shows you intend to both hear and recognize another's perspective. Using your own words, repeat what the speaker has said. By doing this, you'll show you have processed their words, and they will realize your responses have been genuinely thought out. Others will feel more connected to you knowing you are an active listener, and you will develop a better understanding of them.

Tip #5: Promote togetherness

Help others thrive by creating a friendly, cooperative environment and treating everyone the same - not like they're part of a hierarchy, and don't act like one person's opinion is more important than another's. Avoid gossip but always take into consideration the suggestions of others. After addressing a crowd, make sure you've been understood.

If you follow these suggestions, others will come to identify you as a team player and someone who can be trusted.

Tip #6: Resolve disputes

Be the person to whom others can turn when disputes arise. Disagreements can impact everyone - even those not directly involved in the disagreement. You can improve the situation by acting as moderator. Arrange to have a discussion with both of the aggrieved parties and assist them in resolving their conflict. Not only will others be happier, but you will be thought of as a leader.

Tip #7: Be a great communicator

In addition to being an active listener, you need to have other great communication skills. When in a discussion with others, don't blurt out the first thing that comes to mind. Instead, think carefully about the words you use. With clear communication, you'll be able to avoid any potential misunderstandings with others.

A good speaker comes to be known as intelligent and mature, no matter their age. If you have a tendency to give voice to any half-baked thought that crosses your mind, people won't place great value on your opinions or character.

Tip #8: Make them laugh

If you have a great funny bone, use it. As long as you avoid inappropriate jokes and don't laugh off serious situations, you'll find others will be drawn to you. Humor can even be a great way to break down barriers with that super shy or overbearing person.

Tip #9: Put yourself in their shoes

An empathetic person can understand how another person feels, and empathy is an important trait when relating to others. Always consider circumstances from another person's viewpoint. What may seem like the obvious, correct answer to you could have entirely different implications when seen from another perspective.

Above all, keep tabs on your own feelings; people who are unable to tap into their own emotions often have difficulty empathizing with others.

Tip #10: Don't be a whiner or complainer

A chronic complainer is typically the least popular and least liked person in any setting. If you constantly whine about this and that, your negativity will push others away from you. If there's something you really need to get off your chest, write about it in your journal or discuss it with someone who can provide sound advice; then take the appropriate steps to either get a positive/productive attitude, improve the situation or remove yourself from the situation entirely. Otherwise, you will be considered a brat and someone who gives off negative energy.

Human Relationship Skills

Below is a list of human relationship skills important to your success.

- Sensitivity to others
- Treating people fairly
- Listening intently
- Communicating warmth
- Establishing rapport
- Understanding human behavior
- Empathy
- Tactfulness
- Cooperative team member
- Avoiding stereotyping people
- Feeling comfortable with different kinds of people
- Fun person to work with
- Treating others as equals
- Dealing effectively with conflict
- Helping clarify misunderstandings
- Creating an environment of social interaction

Winning Friends and Influencing People

How to Win Friends and Influence People is one of the first best-selling human relationship skills books ever published. Written by Dale Carnegie (1888-1955) and first published in 1936, it has sold over 30 million copies world-wide and gone on to be named #19 on *Time Magazine's* list of 100 most influential books. Business Insider has the book listed as one of the 25 most influential books ever written about business.

Fundamental Techniques for Interacting with People

1. **Don't criticize, condemn, or complain.** Human nature does not like to admit fault. When people are criticized or humiliated, they rarely respond well and will often become defensive and resent their critic. To handle people well, we must never criticize, condemn or complain, because not only is it not the mature way to act, it will never result in the behavior we desire.

2. **Give honest and sincere appreciation.** Appreciation is one of the most powerful tools in the world. People will rarely work at their maximum potential under criticism, but honest appreciation brings out their best. Appreciation, though, is not simple flattery, it must be sincere, meaningful and shown with love.

3. **Arouse in the other person an eager want.** To get what we want from another person, we must forget our own perspective and begin to see things from his/her point of view. When we can combine our desires with another person's wants, he/she becomes eager to work with us, allowing us to mutually achieve our objectives.

Six Ways to Make People Like You

1. **Become genuinely interested in other people.** "You can make more friends in two months by being interested in them, than in two years by making them interested in you."[4] The only way to make quality, lasting friendships is to learn to be genuinely interested in other people and their interests.

2. **Smile.** Happiness does not depend on outside circumstances but rather on inward attitudes. Smiles are free to give and have an amazing ability to make others feel wonderful. Smile in everything you do.

3. **Remember a person's name is, to that person, the sweetest and most important sound in any language.** "The average person is more interested in their own name than in all the other names in the world put together."[5] People love their names so much they will often donate large amounts of money just to have a building named after them. We can make people feel extremely valued and important by remembering their names.

4. **Be a good listener. Encourage others to talk about themselves.** The easiest way to become a good conversationalist is to become a good listener. To be a good listener, we must actually care about what people have to say. Many times people don't want an entertaining conversation partner; they just want someone who will listen to them.

5. **Talk in terms of the other person's interest.** The royal road to a person's heart is to talk about the things he or she treasures most. If we talk to people about what they are interested in, they will feel valued and value us in return.

6. **Make the other person feel important – and do it sincerely.** The golden rule is to treat other people how we would like to be treated. We love to feel important and so does everyone else. People will talk to us for hours, if we allow them to talk about themselves. If we can make people feel

important in a sincere and appreciative way, then we will win all the friends we could ever want.

Twelve Ways to Win People to Your Way of Thinking

1. **The only way to get the best of an argument is to avoid it.** When we argue with someone, no matter if we win or lose the argument, we still lose. The other person will either feel humiliated or strengthened and will only seek to bolster his/her own position. We must try to avoid arguments whenever we can.

2. **Show respect for the other person's opinions. Never say "You're wrong."** We must never tell people flat out that they are wrong. It will only serve to offend them and insult their pride. No one likes to be humiliated; we must handle issues with tact.

3. **If you're wrong, admit it quickly and emphatically.** When we are wrong, we should admit it immediately. When we admit that we are wrong, people trust us and begin to sympathize with our way of thinking.

4. **Begin in a friendly way.** "A drop of honey can catch more flies than a gallon of gall."[6] If we begin our interactions with others in a friendly way, people will be more receptive. Even if we are greatly upset, we must be friendly to influence people to our way of thinking.

5. **Start with questions to which the other person will answer yes.** Do not begin by emphasizing the aspects in which we and the other person differ. Begin by emphasizing and continuing to emphasize the things on which we agree. When people are started in the affirmative direction, they will often follow readily. Never tell someone they are wrong, but rather lead them where we would like them to go with questions to which they will answer "yes."

6. **Let the other person do a great deal of the talking.** People do not like listening to us boast; they enjoy doing the talking themselves. Let them rationalize and talk about the idea, because it will taste much sweeter coming from their own mouth.

7. **Let the other person feel the idea is his or hers.** People inherently like ideas they come to on their own better than those handed to them on a platter. Ideas can best be carried out by allowing others to think they came up with them on their own.

8. **Try honestly to see things from the other person's point of view.** Other people may often be wrong, but we cannot condemn them for that. We must seek to understand them. Success in dealing with people requires a sympathetic grasp of the other person's viewpoint.

9. **Be sympathetic with the other person's ideas and desires.** People hunger for sympathy. They want us to recognize all they desire and feel. If we can sympathize with others, they will appreciate our side as well and will often come around to our way of thinking.

10. **Appeal to the nobler motives.** Everyone likes to be glorious in their own eyes. People believe they do things for noble and moral reasons. If we can appeal to others' noble motives, we can successfully convince them to follow our ideas.

11. **Dramatize your ideas.** In this fast-paced world, simply stating a truth often isn't enough. The truth must be made vivid, interesting, and dramatic. Television has been doing it for years. Sometimes ideas alone are not enough, and we must dramatize them.

12. **Throw down a challenge.** The thing that most motivates people is the game. Everyone desires to excel and prove their worth. If we want someone to do something, we must give them a challenge, and they will often rise to meet it.

Be a Leader: How to Change People without Giving Offense or Arousing Resentment

1. **Begin with praise and honest appreciation.** People will do things begrudgingly due to criticism or for an iron-fisted leader, but they will work wonders when they are praised and appreciated.

2. **Call attention to people's mistakes indirectly.** No one likes to make mistakes, especially in front of others. Scolding and blaming only serves to humiliate. If we subtly and indirectly show people their mistakes, they will appreciate us and be more likely to improve their work.

3. **Talk about your own mistakes before criticizing the other person.** When something goes wrong, taking responsibility can help win others to your side. People do not like to shoulder all the blame and taking credit for mistakes helps to remove the sting from our critique of others.

4. **Ask questions instead of giving direct orders.** No one likes to take orders. If we offer suggestions, rather than orders, it will boost the other person's confidence and allow him/her to learn quickly from mistakes.

5. **Let the other person save face.** Nothing diminishes the dignity of a man quite like an insult to his pride. If we don't condemn our employees in front of others and do allow them to save face, they will be motivated to do better in the future and have confidence that they can.

6. **Praise every improvement.** People love to receive praise and admiration. If we truly want someone to improve at something, we must praise their every advance. "Abilities wither under criticism, they blossom under encouragement."[7]

7. **Give the other person a fine reputation to live up to.** If we give people a great reputation to live up to, they will want to embody the characteristics with which we have described them. People will work with vigor and confidence if they believe they can be better and know that we believe they can do great things.

8. **Use encouragement. Make the fault seem easy to correct.** If correcting a fault seems like a momentous task, people will give up and lose heart. But if it appears easy to correct, they will readily jump at the opportunity to improve. If we frame objectives as small and easy improvements, we will see dramatic increases in desire and success in our employees.

9. **Make the other person happy about doing what you suggest.** People will most often respond well when they desire to do the behavior suggested. If we want to influence people and become effective leaders, we must learn to frame our desires in terms of others' desires.

Simple Rules for Making Your Interactions with Others Better

1. Don't nag.
2. Don't criticize.
3. Give honest appreciation.
4. Pay attention to little things.
5. Be courteous.
6. Never stop trying to improve communication.
7. Continue studying how to improve relationships.

Chapter Conclusion

Through cognizant practice and diligent implementation of good interpersonal skills in all your interactions you will be rewarded with better relationships and enjoy much more success in all areas of your life. It does takes effort, but it is well worth it, as the payoff is incredible.

CHAPTER 17

How to be a Good Employee

Importance of Being a Good Employee

While there may be many reasons people want to be good employees, including promotions, better pay, and social status, the most important ones are self-respect and the sense of accomplishment you get after a job well done. When you are a good employee, you feel good about yourself which in turn helps you be an even better employee and person. Doing your job well gives you a sense of accomplishment. Successful employees also motivate those around them which makes the entire work culture more enjoyable and successful.

If you want to be successful at your job and move up the company ladder you ALSO need to be passionate about your work. You need to be motivated and driven to be the best you can be regardless of your job or your work. Passion, drive, motivation, zeal, call it what you want it is the self-driven attitude about your job and your work that can help lead you down the path to success.

What Makes a Good Employee?

There are many characteristics of a good employee, and many of them mirror attributes of being a good person.

1. **Introspective** – be able to critique yourself to a degree. How can you do things better? What do you need to work on?
2. **Open-Minded** – be open minded not only about how to do your job but also about your co-workers and their beliefs and/or lifestyles.
3. **Proactive** - go above and beyond, join committees, take on responsibilities not necessarily in your job description.
4. **Invested** – be fully invested in your job; always have an answer, opinion, etc. when asked questions on the spot.
5. **Advocate** – not only for yourself but for your co-workers as well.
6. **Genuine** – be honest when you make mistakes, when you need help, and in your relationships with your co-workers and bosses.
7. **Passionate** – love what you do. if you're stuck in an unfulfilling career, you can switch jobs or switch beliefs.
8. **Courageous** – don't be afraid to fail.
9. **Motivational** – use your skills to motivate your colleagues.
10. **Capable** – don't just do your job, do it well and do it with pride.

11. **Focused** - keep personal calls, emails, texts, etc. to a minimum on the job.

12. **Professional** - follow rules, be courteous, friendly, and tactful when dealing with clients, co-workers and bosses; answer emails and phone calls in a timely manner; meet deadlines.

13. **Positive** – have a genuinely positive attitude in the workplace.

14. **Dress for success** – dress appropriately for the job.

15. **Integrity** – integrity is perhaps the most important principle of being a good employee and being an effective leader. Success is dependent on integrity because it demands truthfulness and honesty. Many employees, leaders, companies, and organizations fail because they compromise when it comes to integrity. Integrity means telling the truth even if the truth is ugly. Integrity means doing the right thing at all times and in all circumstances, whether or not anyone is watching. It takes having the courage to do the right thing, no matter what the consequences will be. Building a reputation of integrity takes years, but it takes only a second to lose, so never allow yourself to ever do anything that would damage your integrity.

The value of the trust others have in you is far beyond anything that can be measured. For entrepreneurs it means investors that are willing to trust them with their money. For employees it means a manager or a boss that is willing to trust them with additional responsibility and growth opportunities. For companies it means customers that trust giving them more and more business. For you it means having an army of people that are willing to go the extra mile to help you because they know that recommending you to others will never bring damage to their own reputation of integrity. The value of the trust others have in you goes beyond anything that can be measured because it brings along with it limitless opportunities and endless possibilities.

Tips for Loving Your Career and Working with Passion

1. **Know what you love** *before* you enter the workforce or accept a job. Before you take steps down a specific career path and get a degree in a specialized area, make it your mission to figure out what it is you are passionate about.

2. **Identify what you value in life.** Is it achievement, security or purpose? Choosing a career you love and living with passion begins with understanding what it is you value," Sanders says. "Do you want to have plenty of time for family and friends? Do you prefer working independently or as part of a bigger team? Once you've taken the time to understand your own priorities and talents, then do your research and have the courage to move in that direction.

3. **Find the interaction between what you love and where you excel.** Build happiness into your career by finding the interaction of what you love and where you excel professionally. Work doesn't feel like work when you enjoy what you're doing and see that you're making a positive difference.

4. **If you say you'll do it, do it.** Doing what you love and feeling passionate about your work only goes so far. Back up that passion with characteristics that will get you farther. Integrity is everything. If you commit to something, follow through on that promise.

5. **Trust in the power of attraction.** When you do what you love, people rally round you. Enthusiasm

is contagious and you attract other people who are passionate too.

Tips to Being a Good Employee

There are also things you can do at the office to ensure you are the best employee possible and be successful.

- **Take initiative** – find better, more efficient ways to do your job
- **Be a team player** – work well with others, help others when they need it, ask for help when you need it, and give credit to whom earned it
- **Get to know your boss** – know how he/she thinks, acts, and manages. Does he/she prefer written or verbal updates? Does your boss want updates in an email, during staff meetings, in one-on-one meetings, or on a voicemail?
- **Understand your boss' goals and/or responsibilities** – what is he/she accountable for? Help your boss be successful by being prepared with information.
- **Never let your boss be blindsided** – no surprises.
- **Don't expect your boss to hold your hand** – find a colleague to help you learn where office supplies are kept, how to file daily reports, how to deal with HR, insurance claims, etc. or take the initiative to learn on your own, but don't bother your boss with these.
- **Meet your deadlines** – better yet beat them, giving your boss extra time in case of new developments.
- **Handle constructive criticism** – use critiques to better your performance, don't get your feelings hurt.
- **Cultivate relationships** – workplace friendships help create a positive atmosphere and motivate you and the people around you to do your jobs well.
- **Be willing to learn new skills** – attend training, learn other jobs in your department. It increases your value.
- **Always be part of the solution, not the problem** – being a problem solver indicates you care about the business. Never complain or talk to your boss about a problem without having a possible solution in mind.
- **Be reliable** – if you say you will do something, do it.
- **Avoid gossip** – don't participate or incite office gossip and/or rumors.
- **Volunteer** - for projects, committees, teams, etc.
- **Mentor** - new employees and younger workers.
- **Ask questions** – especially when new on a job, ask until you understand. Don't dig a hole so deep you can't get out or make a mistake that will take valuable time to fix, when a simple question or two could have avoided the situation.
- **Be on time** – always be punctual in beginning your work day.
- **Stay calm under pressure** – even when you feel stress, try not to let it show.
- **Don't worry about perfection** – perfection is not a realistic goal. Mistakes will happen, just be sure you deal with them honestly and quickly.
- **Know what allows you to do your best work** – quiet hour in the morning, protected time during the day with no phone calls or interruptions, etc.
- **Respect your colleagues**
- **Understand the business/organization** – Be familiar with the organizational chart and mission statement. Know the goals and objectives.
- **Focus on bringing value to your employer** – Whether it be through greater revenue, higher profit margin, lowering overhead, learning how to do something your employer needs, or improving efficiency always seek to increase your value and make yourself indispensible.

On a more personal note, you should be sure to get the amount of sleep each night you require to be at your best each day. You should take breaks during the day to stay fresh and focused. Exercise and diet will also help with this. Lastly, don't get caught up in the "busy" syndrome. Some people want to complain or brag about being so busy all the time, they really don't get much accomplished. Instead, be proud of being efficient and on top of your assignments and projects.

Steps to Help Advance Your Career

If you want to move up within your organization and beyond, here are a few tips to help you do so.

- Think about skills you need for your next job – don't be content with knowing your duties, learn what skills and duties are required at the next level
- Speak up in meetings – don't be intimidated in big meetings or when executives are present. Let them know you have something to contribute.
- Get to know management above you
- Take charge - when there is a problem in your office be the one to fix it or get it addressed

- Look for leadership opportunities – volunteer to be the lead on projects, mentor new employees, etc.
- Make connections across the organization or business – get to know everyone: janitors, mail room clerks, receptionists, peers, supervisors, executives, etc. and be sure to treat all with respect.
- Always be professional – meet deadlines, answer emails and return phone calls in a timely manner, and DO NOT participate in gossip.
- Listen – always give people your undivided attention to let them know you care about them and help them be motivated.
- Think like a manager – determine what needs to be done next to help the team succeed and ensure it gets done.
- Keep a record of your accomplishments – be prepared to share these at the right times and in an appropriate manner.
- Know and focus on specific results – be able to give facts and statistics involving projects
- Know who else is getting promoted and why – see what it takes to get promoted at your business.
- Don't compare yourself to others – You should pay attention to others getting promoted but don't change who you are. Know what you uniquely bring to the table.
- Identify ways to make the workplace better and do your part to make it happen.
- Don't avoid business politics completely – know the unofficial rules of your office and who makes things happen.
- Be sure your body language projects leadership qualities – demonstrate control, knowledge, and power by using confident body language such as straight shoulders, chin up, comfortably wide stance when standing, etc.

- Appear comfortable under pressure – "never let them see you sweat."
- Look for opportunities to present at conferences or in big meetings – put yourself "out there"
- Ask supervisors how to get where you want to be – let your boss know where you would like to be in the next few years and let him/her help you.

Being More Successful and Enjoy Your Work

What is success at work? How do you know when you have achieved it? What does success mean to you? Why should that matter? It matters what you think because how you define success will help determine whether or not you achieve it. Success at work for most people means:

- Having a job.
- Doing it well.
- Getting satisfaction from it.

If you are to succeed, you must accept the responsibility for that success. Accepting that responsibility means:

- Setting goals/making plans.
- Taking the initiative/don't wait to be told.
- Making good choices/use good judgment.
- Making no excuses.
- Admitting your mistakes.
- Making your own decisions.
- Solving your own problems.

Conclusion

Change how you think. Like your job. These are the two most important aspects of being a good employee. It is next to impossible to exhibit the characteristics listed above or perform the suggested tips, if you don't genuinely feel that way or like your job. If you don't particularly look forward to going to work every day, start looking at things differently. Instead of dreading seeing the one person in the office that drives you crazy, focus on getting to see the people you genuinely like. Most issues at work are triggered by people, not the job itself. Change how you think about the people, and you will be able to change how you think about the job. If you think your job is boring or beneath you, examine these thoughts closely and determine why you think this. Tasks are neither boring nor exciting; we make them what we want them to be.

CHAPTER 18

Interdependency

What is interdependency?

Interdependency occurs when two or more entities or people rely or depend on one another for mutual benefit. This involves building and developing mutually supportive relationships in both our professional and personal lives to reach our maximum potential.

Stephen Covey, in his book *The 7 Habits of Highly Effective People*, states that "to be able to realize our full potential and reap the rewards, we need others." He discusses the stages of our lives beginning with a state of dependence, when we are taken care of by others; next, is the state of independence when we become self-sufficient; finally, some people have the maturity to reach interdependence. Covey states, "Interdependence is a far more mature, more advanced concept. If I am interdependent, I am self-reliant and capable, but I also realize that you and I working together can accomplish far more than, even at my best, I could accomplish alone....Dependent people cannot choose to become interdependent. They don't have the character to do it; they don't own enough of themselves."

Why is Interdependence Important for your Business?

The activities involved in business are typically interdependent by nature. The basic business tenets of developing an idea, marketing, sales, delivery and finance are all mutually reliant upon each other for the business to succeed as discussed in other chapters. You may have the best idea in the world, but if you don't have marketing to let people know about your service/product, then your great idea is useless. Likewise, if you have a great marketing scheme but don't deliver the service/product as advertised, you will not succeed. Therefore, interdependence is inherent to the success of your business.

Every business goes through the same basic growing stages as a person. When starting a business, you are basically in a dependent or fledgling stage, relying on others for financing, ideas, etc. You try to make your business more independent by breaking away from others and growing it to stand on its own. However, businesses cannot succeed by standing on their own. They will always be dependent on suppliers, customers, employees, etc.

The real success in business comes after the survival and stability stage (independence). This is when the business is ready to move into interdependence. Success comes when customers are dependent on your business, and your business builds and develops relationships with other businesses, resulting in mutual benefits.

Basic Types of Interdependence within a Business

Every business relies on workflow interdependence to some extent as departments, employees, and/or functions rely on each other to achieve the intended outcome. The structure of that interdependence depends on the desired outcome and how much interaction and communication is needed to achieve it. There are three basic types of interdependence:

1. Pooled Interdependence – this is the loosest form of interdependence and more commonly found in large businesses or corporations as opposed to small businesses. In pooled interdependence each department or unit performs separate functions with little communication between them. Their overall goal or product is the same; therefore, they are loosely interdependent upon each other.

2. Sequential Interdependence – each step in the process, as the name suggests, is dependent upon the previous step. Workflow is a step-by-step process in which a delay in one step delays future steps. Sequential interdependence requires more communication than pooled interdependence. This type of interdependence is most commonly used in manufacturing, construction, and financial industries.

3. Reciprocal Interdependence – this is the most stringent form of interdependence. Each step in the process is dependent on the previous one, as in sequential interdependence; however, in reciprocal interdependence it is also cyclical. The dependence goes in both directions: forward and backward. This requires the highest level of communication between the steps in the workflow. No single area or department can complete a task or make a decision without input from the whole. Examples of businesses typically employing

reciprocal interdependence include restaurants and hotels.

How to Ensure Your Business is Employing Interdependence?

Employing interdependence means working with others to achieve a goal. Interdependence in the workplace includes how employees interact with each other and how individual skill sets are used to the maximum capacity: Working together on a project that cannot be completed by one person ensures the project is completed, while allowing for individual input on group projects ensures the best possible outcomes. Therefore, leadership is the key to successfully employing interdependence in the workplace.

- **Develop leadership as opposed to leaders** - Studies show there is an emerging trend away from training leaders to developing leadership. The *Harvard Business Review*, in its article titled "A Declaration of Interdependence," explained companies "…have expanded from an exclusive focus on leader development, which is about character, competence, quantity of individuals in defined roles, to leader*ship* development which is the expansion of a collective's shared beliefs and practices for creating direction, alignment, and commitment (DAC)."
  - ◊ **Direction** – How will we decide where to go from here?
  - ◊ **Alignment** – How will we coordinate our work?
  - ◊ **Commitment** – How will we stay engaged and accountable?
- **Delegate** - Supervisors should delegate tasks to others who can apply more expertise, attention and/or time than the supervisor himself has available. If a supervisor does not delegate, he runs the risk of missing deadlines or producing poor quality work, both of which can be detrimental to the business.

- **Acknowledge interdependent successes** - Leaders should also make it a point to acknowledge successes from interdependent efforts. It is always easy to recognize individuals for feats such as "top sales;" however, it is just as, if not more so, important to recognize when goals are achieved as a result of a team effort.
- **Foster and develop connections** - When people care about each other, they tend to share at least some of the same values, and this leads to a shared motivation when working on a project together. Supervisors should foster and develop connections among their employees. This can be done by emphasizing and demonstrating how success is dependent on working together and helping each other.

Summary

Everything and everyone is connected in some way, and we are interdependent on many things in both our personal and professional lives, sometimes without even realizing it. No one and no business can succeed by being entirely dependent or entirely independent; real success only comes through interdependency.

Attempting to run a business independently is self-defeating. Each business needs other businesses and needs customers and other businesses to need it. Employees and/or processes within a business are interdependent on other employees and/or processes to achieve goals and objectives. So, we should utilize this interdependency to our advantage to achieve maximum success.

Humans typically want to feel in control of their lives, so it is important to point out the things we do control. We control the reason our business exists. We control how we run our business. We control our motivation. We control how we compete with other businesses. We control the quality of our products. The key is to realize and understand that we do not have this control on an island independently. We are interdependent on other people and businesses to succeed.

PART III

Taking Success to the Next Level

CHAPTER 19

Customer Service Recommendations and Rewards

Customer service relates to the service provided to customers before, during and after a purchase.

No matter the size of your business, excellent customer service needs to be at the heart of your business model for you to be successful. It is important to provide good customer service to all types of customers — potential, new and existing customers.

Although it can take extra resources, time and money, excellent customer service can generate positive word-of-mouth promotion of your business, keep your customers happy and encourage them to purchase from your business again. Good customer service can help your business grow and prosper.

Significance

Customer service is important to an organization, because it is often the only contact a customer has with a company. Customers are vital to an organization. Some customers spend hundreds and even thousands of dollars per year with a company; consequently, when they have a question or product issue, they expect a company's customer service department to resolve that issue.

Identification

Customer service is also important to an organization, because it can help differentiate a company from its competitors, according to the article titled "The Importance of Customer Service" on the Drew Stevens Consulting website. For example, it may be difficult for consumers to choose between two small-town drug stores, especially if their prices are similar. Therefore, putting extra efforts into customer service may be thing that gives one drug store a competitive advantage.

☑ AWESOME!

☐ **Excellent**

☐ **Very Good**

☐ **Satisfactory**

☐ **Marginal**

☐ **Poor**

Function

A company with excellent customer service is more likely to get repeat business from customers. Consequently, the company will benefit with greater sales and profits. Contrarily, companies with poor customer service may lose customers, which will have a negative impact on business. It costs a lot more money to acquire customers than to retain them, due to advertising costs and the expense of sales calls. Therefore, the efforts that go into maintaining quality customer service can really pay dividends over time.

Publicity

People that have a positive experience with a company's customer service department will likely tell two or three others about their experience, according to the Consumer Affairs website. Therefore, quality customer service can be a source of promotion for organizations. Contrarily, a person who has a bad customer service experience will likely tell between nine and 20 people.

Prevention/Solution

Customer service is important to an organization because of potential complaints. Consumers can file a complaint with the Better Business Bureau, Consumer Affairs or even a class action attorney if they are dissatisfied with a company's customer service. Legally, consumers are protected by the Federal Trade Commission and can sue a company in the courts. Equally as damaging, if not more so, are the negative reviews e and/or rants through social media a customer can post online.

Specifics as to Why Customer Service is so Important

It can help you
- increase customer loyalty
- increase the amount of money each customer spends with your business
- increase how often a customer buys from you
- generate positive word-of-mouth and reputation
- decrease barriers to buying (for example, if your business has an excellent reputation of customer service for refunds, you're more likely to entice a hesitant buyer to purchase from you).

Excellent Customer Service Defined

Excellent customer service is about
- treating your customers respectfully
- following up on feedback
- handling complaints and returns gracefully
- understanding your customers' needs and wants
- exceeding customer expectations
- going out of your way to help your customers.

Types of Customer Service

Issue-Centric Customer Service

Customer service models that focus on addressing issues rather than serving people can be classified as 'issue-centric customer service.' From the company's perspective, an issue is (1) a subject line, (2) a brief description of the problem reported by the customer, (3) the customer's email address, and (4) the issue's status and priority.

Product-Driven Customer Service

With a product-driven company all functions are focused on the product — its design, features, capabilities, and its subsequent design and manufacture. Customer service operates under the assumption that putting a strong emphasis on great products will result in higher profits and revenue.

Customer-driven customer service

A Customer-driven customer service model revolves around people. Every component of the solution is humanized, acknowledging that customers are real people with daily life struggles like the rest of us. It is a model that recognizes customers as a valuable resource, key to a company's success; therefore, this model is built to respect customers.

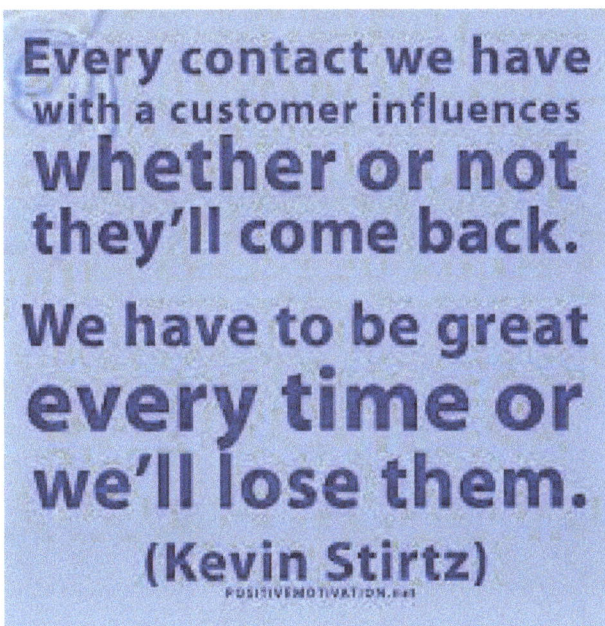

Every contact we have with a customer influences **whether or not they'll come back.** We have to be great **every time or we'll lose them.** (Kevin Stirtz)

Developing and Maintaining Excellent Customer Service

There are certain customer service skills every business (management and employees) must master, if they are to achieve optimum success. Without these skills business growth, revenue, and profits will be limited, and customer-related problems will be amplified.

Fortunately, there are a few universal skills that all businesses (management and employees) can master to dramatically improve their relations with customers, which in turn produces positive results. Below are the 17 most-needed customer service skills to master.

The Customer Service Skills that Matter

When most business publications talk about customer service skills, things like "being a people person" tends to take the spotlight. Although such a statement is not exactly wrong, it is so vague and generic that it is of little help. The following are specific practical skills that management and employees can realistically implement to "WOW" the customers, gain loyalty, and grow their business.

1. Patience

Not only is patience important to customers, who often reach out for support when they are confused and frustrated, but it's also important to the business at large. Great service beats fast service every single time. Yet patience shouldn't be used as an excuse for slow service.

Quality time spent with the customer to better understand his/her problems and needs is an essential component of providing excellent customer service. It is imperative to remain patient with customers and take the time to truly figure out what they want, when they come to you stumped and frustrated.

2. Attentiveness

The ability to *really* listen to customers is crucial for providing great service. Not only is it important to pay attention to individual customer interactions (watching the language/terms they use to describe their problems), but it's also important to be mindful and attentive to the feedback you receive *at large*.

For instance, customers may not be saying it outright, but perhaps there is a pervasive feeling that your product or service isn't as great as it could be. Customers aren't likely to say, "Please improve your website;" however, they may say things like, "I can never find the search feature," or, "I am having trouble pacing and order." Listening to your customers can help you focus on things that truly make a difference in improving your business, products, and/or service.

3. Clear Communication Skills

It is very important to remain professional and cautious in your communications. Customers don't need your life story, your opinion on issues, or to hear how your day is going. It is always best to err on the side of caution whenever you find yourself questioning a situation. Positive results from excellent customer service are achieved through simple, concise communications, leaving nothing to doubt.

4. Knowledge of the Product and/or Service

Businesses that have the best customer service ensure all employees interacting with customers have a clear understanding of the product and/or services being offered. It's not that every team member should know everything about the product or service, but rather they need to know the functions of the product or service being offered, just like a customer who uses it every day would. Without such knowledge it is impossible for these employees to help customers efficiently, when they run into problems.

5. Ability to Use "Positive Language"

Sounds like fluffy nonsense but having the ability to make minor changes in conversational patterns can truly go a long way in creating happy customers.

Language is a very important part of persuasion, and people (especially customers) create perceptions about a person and a company based on the language used by company representatives.

Small changes that utilize "positive language" can greatly affect a customer's perception of a company. Following is an example of positive language versus not-so-positive language.

- Without **positive language:** "I can't get you that product until next month; it is back-ordered and unavailable at this time."
- With **positive language:** "That product will be available next month. I can place the order for you right now and make sure it is sent to you as soon as it reaches our warehouse."

The first example isn't necessarily *negative*, but the tone it conveys feels abrupt and impersonal, and the representative could be taken the wrong way by customers. Conversely, the second example is stating the same thing (the item is unavailable), but this example positively focuses on when/how the customer will get his/her resolution.

6. Acting Skills

It is impossible to please everyone all of the time. Some situations are outside our control. The other person may be having a terrible day or may just be a natural-born complainer. Some people seem to want nothing else but to pull others down.

A polished customer service representative has the *basic acting skills* necessary to maintain a cheery persona in spite of dealing with people who may be just plain grumpy. Regardless of another's words, tone of voice, or complaints, responding in a polite cheery way exuberates excellent customer service and does much to separate outstanding companies from average companies.

7. Time Management Skills

Despite many publications stating the importance of spending more time with customers, there *is* a limit to how much time you can afford to give a customer. It is also important to be respectful of the other person's time. Therefore, the key is to provide customers what they want in the most efficient way possible.

When a company representative can no longer help or is not making any progress with a customer, it is time for the customer to be connected to someone within the company that can assist him/her further. It is critical to not waste time trying to go above and beyond for a customer when it is not a prudent use of time and resources (where the cost outweighs the benefit).

8. Ability to "Read" Customers

Many customer interactions are not face-to-face. Additionally, because so much business is conducted through the Internet there is often not even a verbal interaction. However, there are some basic principles of behavioral psychology (being able to "read" the customer's current emotional state) that still apply.

This is an important part of the personalization process as well, because it still means understanding customers in order to create a personal experience for them. This skill is *essential,* because misreading a customer and losing that customer due to confusion and miscommunication is costly. To best achieve positive customer interactions, search for subtle clues about the customer's mood, patience level, and attitude in order to respond in appropriate and inviting ways.

9. A Calming Presence

There are several metaphors for this type of personality: "keeps their cool," "stays cool under pressure," etc., but they all represent the same thing. These people have the ability to stay calm and even influence others when things get hectic. The best customer service representatives know they *cannot* let a heated customer force them to lose their cool; in fact it is

their *job* to try to be the "rock" for a customer who thinks the world is crumbling due to his/her current problem. Responding to a difficult customer with a negative comment or wrong attitude will, at a minimum, result in a lost customer and, at worst, agitate a customer to the point he or she will leave negative reviews and bad mouth your company to others.

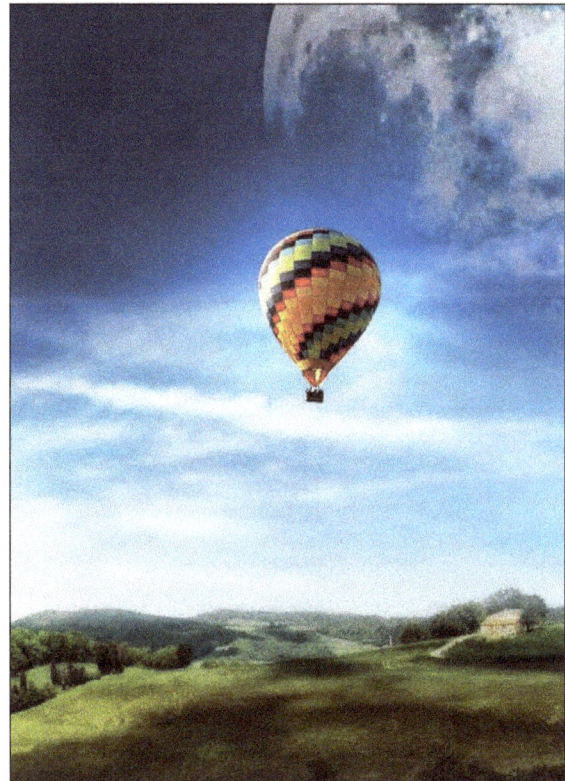

10. Goal Oriented Focus

This may seem a strange thing to list as a customer service skill, but it is vitally important. Several business studies have shown how giving employees' unfettered power to "WOW" customers isn't perceived by customers as good service. Furthermore, the more relaxed or liberal customer service approach doesn't always generate the returns companies initially projected it would.

Rather, relying on customer service frameworks and goals are what help a company achieve the desired results. Instilling structured customer service guidelines — but allowing some freedom to handle customers on a case-to-case basis — efficiently helps customers resolve problems and produces much better results.

11. Ability to Handle Surprises

Whenever interacting with customers the rule of thumb should always be to expect the unexpected. Maybe the problem encountered isn't specifically covered in the company's typical protocol, or maybe the customer isn't reacting as anticipated. Whatever the case, it's best to create guidelines that will best address these situations.

For these instances it is best to create a 'fallback' plan. Following are some examples.

- **Who?** Define who the "go-to" person will be for issues not covered in the company's policies and general practices. Define a logical chain of authority to address problems and answer unusual questions.
- **What?** Define *what* issues will be sent to the people above.
- **How?** Determine the methods as to *how* employees are going to contact those on the chain.

12. Persuasion Skills

The power of persuasion is of extraordinary and critical importance in today's world. Nearly every human encounter includes an attempt to gain influence or persuade others to our way of thinking. In business, it can mean the difference between success and failure — sustainable revenue and extraordinary profits.

For many, the notion of becoming a polished persuader means being forceful, manipulative, and/or pushy. Such an assumption is absolutely wrong. Tactics like these may get short-term results, but Maximum Influence is about getting long-term results. Maximum influence isn't derived from calculated maneuvers, deliberate tactics, or intimidation. Rather, proper implementation of the latest persuasion strategies will allow one to influence with the utmost integrity.

Maximum Influence supplies a complete toolbox of effective persuasion techniques. Most people use the same limited persuasion tools over and over, achieving only temporary, limited, or even undesired results. These tools and techniques are addressed in full detail within the Persuasion chapter of this book.

13. Tenacity

Tenacity in business — more specifically in 'customer service' — involves having a great work ethic and a willingness to do what needs to be done. It is a key skill needed to provide the kind of service people talk about — in a positive way.

The many memorable customer service stories out there (many of which had a huge impact on the business) were often created by a single employee who refused to just do the 'status quo' when it came to helping someone out. This translates into positive 'word of mouth' referrals (free marketing) and higher profits. Putting in the extra effort generates exponential returns and therefore should be a driving motivator to never be lax or too casual in providing excellent customer service.

14. Closing

The 'closing' being referred to here has nothing to do with 'closing a sale.' Being able to close with a customer means being able to end the conversation with confirmed satisfaction (or as close to it as you can achieve) and with the customer feeling that everything has been taken care of (or will be).

It is important to take the time to confirm with customers that each and every issue he/she had on deck has been entirely resolved.

Properly closing shows the customer four very important things:

- That you care about getting it right.
- That you're willing to keep going until you get it right.

- That the customer is the one who determines what "right" is.
- That you care about the customer.

A happy customer is one who can say, *"Yes, I'm all set!"* following the interaction. Such is a win-win for both parties.

15. Adaptability

Every customer is different, and some may even seem to change week to week. Being able to handle surprises, sense the customer's mood and adapt accordingly is critical.

16. Thick Skin

The ability to swallow one's pride and accept blame or negative feedback is crucial. Whether working directly with customers or looking for feedback on social media, great service is doing everything reasonably possible to assist the customer, regardless of the customer's disposition.

17. Willingness to Learn

This is probably the most general skill on the list, but it is necessary. Those who don't seek to continually improve what they do, whether it's building products, marketing businesses, or helping customers, will get left behind by the people willing to invest their time making improvements.

Importance of Saying 'Thank you'

The simple act of saying "thank you" to customers and clients is all it takes to make your business stand out from the competition. Saying thank you is not just about having good manners it's about staying ahead of competition.

Customers have a choice. In fact, they have more choices than ever before when it comes to where and how they buy products and services. Chances are good that whatever they buy from your organization they could find elsewhere and perhaps even at a lower cost. You should certainly understand why customers buy from you but, more importantly, you should thank them for that decision. Doing so will remind them why they do business with you.

Clients want to be regularly reminded that they are important to you. It's not up to your clients to remember you. It's your job to constantly and consistently reinforce the relationships with your customers.

If you want your customers to buy from you year-round, then your gratitude should extend year-round as well. The No. 1 asset your business has is your client relationships.

Jesus healed ten leapers, and only one returned to say "thank you". The same ratio of gratitude exists today. As you make your next purchases note how rare it is for clerks to say thank you. By simply saying 'thank you' your business will stand out from the competition. It also makes people want to return and do business with you again because they feel appreciated.

Being grateful to others makes them feel like their actions are more meaningful; and people have a basic need to feel appreciated. Those feelings cannot be overstated. The success of your business relies on these happy and loyal clients. They may not always remember what you say, but they will remember how your business makes them feel.

Saying "thank you" may sound like a cliché but it can provide tremendous dividends. It makes the person receiving it feel special and makes the person saying it stand out from the crowd. Saying "thank you" cost nothing but goes miles in making people want to remain connected with your company and its offerings.

27 Creative Ways to Thank Customers or Clients

Saying, "Thank you" is important,, if not critical to optimizing business success. However, just simply saying 'thank you' is often not enough. Some things just demand a little something extra because of the magnitude of the favor or the depth of appreciation involved; and also because of the rewards the gesture can provide. Being more creative, or doing more than just saying 'thank you' or firing off an email is warranted. Following are creative ways to say 'thank you' that others will remember and cherish!

1. Give a Gift

When it comes to thanking customers, it really is the thought that counts. There is no need to offer anything lavish. Choose something your customers will benefit from and is customized in some way. Examples include a book that influenced you, homemade treats, or a gift card. Gifts should just be for your top customers — if you offer one to everyone, it loses its meaning.

2. Send a Handwritten Note

Even if you have a very limited budget, you can thank your customers. For instance, if you have just a handful of loyal customers, send out handwritten notes to everyone. The effort you put into this will mean much more than an email — and it is less likely to go unnoticed. Make each note personal, even if this means you need to send fewer.

3. Hold a Competition

If you have enough in your budget for just one amazing gift, hold a competition with the chance for one customer to win. Give the contest a name that ensures participants are aware you are holding it to thank them.

It is best if you can make the competition revolve around your brand. For example, you could ask customers to tell stories about how one of your products

improved their lives. Use social media to garner more attention for your business.

4. Post Photos

Another way to utilize social media is to post photos of your customers (with their permission) on your pages. As well as allowing you to thank customers, this will show real people interacting with your products and using your services.

5. Make a Charitable Donation

Advertise your brand's values by making charitable donations in the names of your customers. Additional benefits can be obtained by choose a local organization or a charity that has some type of connection to your industry or business.

6. Teach Something New or Provide Solutions

Consider what problems your customers have and offer solutions. The opportunities for distributing your knowledge are endless. For instance, you can provide an ebook, webinar, or infographic or even invite customers to an online knowledge center. If you want to interact with your customers in person, hold an event that gives them the chance to learn something new and useful related to your offerings.

7. Host a Thank-You Event

Another way to bring all your top customers together is to hold an event with the sole purpose of thanking everyone. There is no need to invest much — just a few refreshments are fine. This can also be another opportunity to make additional sales during the event.

If you're not a local company, this idea can still work for you. Invite everyone to a live webinar or Facebook video where you can thank them.

8. Meet With and Talk to Customers

If you'd rather have more personal interactions, schedule one-on-one sessions with a few customers. This is a great opportunity to find out what people like about your company and what you could be doing better. Meet customers for refreshments or invite them to lunch.

9. Begin a Loyalty Program

Add customers to a new loyalty program after they spend a certain amount or visit your business a specific number of times. Rewards can be special offers, discounts, or freebies.

10. Reciprocate with Support for their Business

If some of your customers have their own companies in the area, you can show your appreciation by supporting their businesses. You can even go beyond simply choosing them over their competitors. For instance, you can recommend their products and services in your store, refer your customers to them, mention them on social media, or even reach out about collaborating on a future event.

11. Offer a Complimentary Upgrade

If you provide a service at different levels of quality, surprise a few of your top customers with a complimentary upgrade.

12. Give a Gift Bag

A unique, homemade gift bag with a custom label or a note is a simple but heartfelt way to show your appreciation.

13. Give a Toast

Many people fear public speaking more than death, giving this particular thank-you a little extra meaning. Composing a sincere, eloquent toast and delivering it is a nice way to show appreciation to them in front of others.

14. Write a Poem that is Customer or Client Specific

If you are not the creative type, look for ideas on the Internet. For instance, you can find an applicable poem and switch out some of the words.

15. Create Your Own Labels

There are a number of websites that offer custom gift labels. Find one that fits your personality and that of your customers for a personalized thank-you label.

16. Give a Gift Card

Sometimes choosing what to give a customer can be tough. A gift card is a good way to get around this problem. However, for maximum impact be sure to include a personalized note or card thanking them for their business.

17. Use Social Media to Send a Special Message

Let the whole world know about your appreciation via a social media blast. Use your blog, your Facebook, your Google+ account, and your Twitter to spread the word about why this a customer means so much to you and or your business.

18. Make Your Own Digital Greeting Card

While an email isn't always the best way to go when saying thank you, a digital greeting card that you put time and effort into creating can really brighten someone's day. Make the card customer specific and compose a short message of thanks for their generosity.

19. Make a YouTube Video

Sometimes, actually hearing someone say, "Thank you," can make all the difference. Why not take it a step further and create a special video of thanks for your clients.

20. Deliver Something You Baked, Made, or Picked

Making something yourself is a fun and delightful way to say thank you to someone. Create a sampling of baked goods or a fruit basket and include a special message. Attach a thank-you note or label and surprise customers with the gift of your time and creativity.

21. Make Surprise Gifts for Customers

There's no need to wait until "later" to send a thank-you message. Why not do it at the time? Create little gift packets or bags for your customers with surprises inside. This is a great way to say thanks to the people who do business with you, or attended your event.

22. Send a Flower Basket to Your Clients

Assembling a flower basket with a thank-you note is an excellent way to brighten someone's day and show you appreciate to clients.

23. Take a Picture

Sometimes capturing the moment is the best way to put a smile on someone's face. Have someone take a picture of your clients during a special moment or while opening that surprise package and send the client a copy with a quick but sincere note to say thanks.

24. Repay by Paying it Forward

The best gifts come from the heart, and the best way to repay a gift is to pay it forward. If your client has a special cause they care about or something they believe in passionately, consider making a donation in their name or volunteering some of your time to the cause.

25. Do Something Special

Take them out to dinner or a movie. Cook them dinner and give them a present when they arrive. Any of these are good options for showing a client you really appreciate them and how grateful you are for their business.

26. Reciprocate with Help or a Kind Gesture

Everyone needs help sometime. Whether it's holding their hand through a particularly traumatic incident or helping them through car trouble. Sometimes just giving them a place to come hang out when they're lonely or showing up to offer them a sympathetic shoulder means the world to a person. Being there when they need it shows you truly care about them.

27. Listen to Your Clients and Customers

Listening is almost as lost an art as the handwritten letter. When customer or clients needs to talk, listen to them. Ask questions when appropriate, but just letting them know you're there and paying attention to them is a great way to say 'thank you' and show you care about them. Offer to pray for them, and then do pray for them. Pray for them while with them and when you are on your own.

CHAPTER 20

Marketing

What is Marketing?

The Merriam-Webster dictionary states marketing is "the process or technique of promoting, selling, and distributing a product or service." In other words marketing encompasses almost every aspect of your business from introducing your product to potential buyers to placing it in their hands, then maintaining a positive relationship with your client to ensure they come back for more products.

You may be the "best in the business" or have the best product available, but if no one knows about you or your business, you won't succeed. To ensure you and your business do succeed, you need a good marketing strategy.

The 7 Ps

A good marketing strategy takes into considerations the 7 Ps:

1. Product – evaluate and continually re-evaluate your product. Is it the best it can be? Is it relevant to your customers?

2. Price – evaluate and continually re-evaluate the price of your product(s). Is the price appropriate for the current market? Are you making a profit on this product? Is the profit worth the time and effort it takes to produce it?

3. Promotion – continually think in terms of promotion. Are your methods of getting information to current and potential customers working? If not, change them.

4. Place – Is your product in a place where customers or potential customers can easily access it? This could be a physical or online place – preferably both for maximum exposure.

5. Packaging – evaluate and continually re-evaluate your packaging. Packaging includes everything a potential customer sees from first contact with your business (either online or physical) to the closing of the sale.

6. Positioning – evaluate and continually work on improving your positioning in the hearts and minds of your customers and in your market. What do people think about you and your business? What do people say about you and your business? What do people say about your products?

7. People – Do you have people within your business doing what they do best? A friendly/social person answering the phone and/or greeting people when they enter your business? Responsible, ethical people selling your product? Note: if you are starting a small business, you (or your employee(s)) may have to "be" more than one of these people.

Why Should I Market my Business?

The most important reason to market your business is to gain higher sales to make your business a success; however, there are other benefits to marketing that go along with that. Marketing will get the word out to potential customers and make them aware of your product/services. Marketing will help you build and increase the visibility of your brand. It will help build a good reputation for your business. Marketing will develop lasting relationships and develop loyalty and trust with current and prospective customers. The following is a list of other benefits to marketing your business:

- Builds authority and credibility
- Positions your business as an expert
- Generates traffic to your site which increases leads
- Helps customers move through purchase decisions more quickly
- Constantly reminds customers you are still in business
- Enables you to determine your target audience
- Provides constant information to your customers
- Allows you to learn about your customers
- Helps you focus on goals and expectations

Types of Marketing

Four common types of marketing include Online Marketing, Traditional Marketing, Relationship Marketing, and Internal Marketing.

Online Marketing

Online marketing, as the name suggests, includes optimizing all online avenues to reach potential customers with information about your business. This type of marketing makes your information available to online and mobile devices allowing potential customers exposure to your business at all times and right in the palm of their hands. Below are a few examples of online marketing:

- **Inbound** – blogs, podcasts, videos, e-books, whitepapers, and online newsletters
- **Email** – delivering your information and promotional offers to people through their email inboxes
- **Ads** – these are banner ads you see when you visit websites; NOTE: recent studies indicate people are beginning to ignore these ads due to the abundance of them, so this may not be a good type of marketing at the beginning
- **Search Engine** – this involves utilizing search engines such as Google, Yahoo, Bing, etc. There are two types of search engine marketing:
  1. organic – optimizing your site with particular words that receive more clicks which results on more visits to your website;
  2. paid – buying ads so the link to your website is more visible in search engines.
- **Mobile** – utilizing push notifications, multimedia messages and text messages that go directly to potential customers' mobile devices
- **Content** – "technique of creating and distributing valuable, relevant and consistent content to attract and acquire a clearly defined audience—with the objective of driving profitable customer action. The key word here is 'valuable,'" according to Forbes. This is the information used in Inbound Marketing discussed above and can be used elsewhere including white papers under Traditional Marketing discussed below.
- **Social Media** – using social networks such as Facebook, Twitter, Instagram, Snapchat, YouTube, etc. to provide information about your business and

to promote it. See the social media chapter in this book for more information on this type of marketing.

- **Affiliate** – rewarding people for referring others to your business

Traditional Marketing

Traditional Marketing utilizes strategies offline to increase brand awareness and increase sales. Some marketing professionals believe traditional marketing is no longer effective; however, this is dependent on the type of business you are marketing and the audience whom you are trying to reach. Examples of traditional marketing are listed below:

- **Broadcast** – using radio and television ads
- **Print** – using ads in magazines and newspapers or on billboards
- **Snail Mail** – developing brochures, letters, postcards, fliers, etc. and sending them through traditional mail
- **Sales** – taking the sale directly to prospective customers' homes physically or over the phone
- **Booths** – purchasing space at tradeshows, conferences, or any large gatherings
- **Referrals** – utilizing current customers to spread the word about your business—this could include rewards for those current customers and the potential customers when they become customers

Relationship Marketing

The focus of Relationship Marketing is to build strong relationships with current customers. This type of marketing helps develop emotional ties for the customer through transparency and trust. The basis for Relationship Marketing is data. You must use data on current customers to learn about them and how to develop the relationship you want them to have with you. They, in turn, will refer others to your business.

Internal Marketing

Internal Marketing utilizes employees as marketing tools. This is basically marketing to your employees. This tests your marketing techniques and increases the reach of your marketing. It tests your marketing techniques, because you can determine the success rate of each technique on your employees. Do they actually buy the product? Internal Marketing also increases the reach of your marketing, as employees then become word-of-mouth and/or referral marketing tools, when they believe in the product and begin telling their friends and families about it.

History of Marketing

Marketing is not a new thing. The ancient Greeks and Romans had booths and stalls at the local markets where they engaged in persuasive marketing practices to entice potential customers to buy from them rather than their competition.

During the Industrial Revolution the production of goods was separated from their immediate consumption for the first time, and mass production began. This led to different ways of managing the distribution of products: marketing.

Marketing has been evolving ever since. The changes in technology and the every-day access potential customers have to it has driven and continues to drive mass changes in marketing techniques. Companies must continue to change their marketing techniques to reach their potential customers, as customers continually change the avenues in which they receive information.

This is not to say traditional marketing no longer works. Word of mouth is still considered one of the strongest marketing tools.

Most Effective Marketing Methods

Research on the most effective marketing methods led to the following list. Not every marketing source had them in the same order; however, almost every source had these in their top 10. Also note that some of these actually combine different types of marketing listed above. Remember, your particular niche and your target audience will determine which method works best for you.

1. **Events:** in-person events can help introduce your business to new customers and help strengthen your relationship with current customers. Examples include open houses, discount nights, social events, etc.
2. **Email:** emails allow you to share business information with potential customers and keep in touch with current customers.
3. **Referrals:** establishing solid relationships with employees and current customers lead to referrals which lead to new customers
4. **Content and Search Engine:** these go together as you optimize your content to increase your search engine rankings which leads more people to your site
5. **Calling:** use data on current customers and that received online from potential customers to determine best potential before calling
6. **Traditional:** this includes television, radio, magazines, newspapers, etc.
7. **Snail Mail:** post cards are successful marketing avenues for some businesses

8. **Public Relations:** press releases and charity sponsorships can help create positive public perception which leads to more customers
9. **Social Media:** these are tools that can help develop personal connections with potential customers—be sure to respond and interact with them in a timely manner
10. **Paid Search:** paying to have your website at the top of search engine searches can lead to more people on your site which can lead to more customers

Challenges to Marketing a Small Business

There are numerous ways to market your business; however, some techniques work better than others for different types of business/products. So, it is important to NOT just choose a technique and run with it. This could cause you undue stress and feelings of failure. It is best to understand the challenges of a

project before going into it. The following list shows some of the biggest challenges small businesses face in marketing:

- Setting purposeful goals
- Determining your target audience
- Focusing on your target audience
- Knowing where to reach your target audience
- Knowing what to produce for marketing
- Knowing what format to use in marketing
- Keeping content fresh

How to Successfully make Marketing Work for You

The most successful marketing strategies combine several specific methods. These, of course, should be based on your specific situation. However, there are several things you should do in your marketing campaign to make it successful:

1. **Know your Product** – Yes, this is your business, and you probably know your product, but be sure you know **everything** there is to know - everything it can do, everything for which it can be used, how it is made, its quality – everything.

2. **Do your Research** – Now that you know everything there is to possibly know about your product, research the industry. You need to know your competition. You need to know what marketing strategies have and haven't worked in your niche. Be sure your research includes the most recent marketing success stories, as what works can change over time. How much money and other resources are required for this successful

marketing? Do you have them or are they available to you?

3. **Know your Audience** – Next, you need to determine to whom you are trying to sell your product. Yes, it would be great, if everyone in the work bought your product, but we know that is not feasible, and there is no need to spend money and resources on potential clients who really aren't "potentials." Once you have determined who is most likely to buy your product, you need to learn more about them. Where do they spend their time? Where do they get information? Where do they spend their time online? With whom do they associate? Where do they live? This will help you decide the best avenues to use to reach them.

4. **Have a Plan** – Now, you know your product; you know your competition and successful marketing techniques that have been utilized in your industry; and you know your audience. Put all this information into a plan that works for you. This should be a step-by-step plan of your marketing strategy, including a calendar of when to execute each part of your plan. For example, if you plan on using emails, prepare and schedule them ahead of time, if possible. If not, mark on your calendar, when they should go out.

5. **Execute your Plan** – This is the fun part! Begin executing your plan, according to your calendar.

6. **Evaluate and Adapt** – You should constantly evaluate your plan and/or the results of your marketing techniques. What's working, as in bringing in more customers, and what is not working? Then adapt your plan or specific techniques. Sometimes just a tweak to the way your present information can make a big difference. You also want to be sure to keep any and all content marketing fresh, timely, useful, and interesting.

Trends in Marketing

As stated above marketing must change with the times to be successful, and there are several emerging trends in marketing for the upcoming years of which business owners should be aware. You may want to start thinking about some of these, and how you may or may not be able to utilize them in your marketing strategy. Below is a list of the most rapidly progressing trends:

- **Marketing Automation** – "software platforms and technologies designed for marketing departments and organizations to more effectively market on multiple channels online (such as email, social media, websites, etc.) and automate repetitive tasks" – Wikipedia
- **Location Based Marketing** – this includes geo-targeting (sending messages to customers in specific regions); geo-fencing (allows businesses to know when customers walk in their door and send them specials or daily offers); beacons (low energy Bluetooth devices to target customers in specific locations such as aisles, entrances, etc.)
- **Virtual Reality** – supplying virtual experiences for potential customers
- **Ephemeral Marketing** – implies the use of messages that are available for only short periods of time; the most common current use of this involves Snapchat

Conclusion

Every business owner wants to be as successful as possible; however, many overlook marketing as an integral part of that success. It is easy to get caught up in the day-to-day operations of your business and wonder why your number of customers or sales isn't increasing. A good marketing strategy can help you with this, and when done correctly, the results are well worth the time, effort, and money you put into it.

CHAPTER 21

Branding

Branding, as discussed in several other chapters in this book, should be an integral part of your marketing strategy. Branding develops recognition of and loyalty to your business when done correctly. This keeps your business in the minds of your clients and potential clients so that when they need a product/service you offer, your business is the first one they think of with favorable feelings. This, in turn, leads them to you for that product/service.

A strong brand will let people know what your business does and how it does it in a simple, concise way that is easy for people to remember. Your brand is also your promise to your customers and potential customers of how you will treat them and what to expect from you. Your brand is everywhere including your website, brochures, flyers, posts on social networks, and your daily interactions with people; therefore, it is a very powerful tool that can help make or break a business.

Benefits of Branding

Regardless of whether you consciously work to develop your brand or not, you have one. Technically, every individual has a brand, i.e., others feel a certain way when they hear an individual's name based on previous interactions with that person or what they have heard from others. The same is true in business; therefore, it is worth the time and effort to develop the brand you want.

Branding will

- Help your business stand out from others;
- Build credibility;
- Increase the value of your product/service;
- Help grow your business.

Developing a Brand

Where do you begin, when you're ready to begin developing your brand? Just like in your marketing strategy, you first need to know who your potential customers are and what you can offer them. Where do your clients and potential clients obtain information about services/products they need? What do they need? What is the one thing you want your customers and potential customers to think of when they hear the name of your business? For instance, do you want your business to be known for the best quality services/products, the fastest service/delivery, or being the most knowledgeable in your niche?

Once you have completed your research and defined exactly how you want to brand your business, you're ready to begin the process of branding:

1. **Develop a logo.** Unless you are a graphic artist, this may be a good place to invest a small amount of money to hire an expert to develop your logo. It will represent your company, so you want it do so professionally. A good logo should be
   - Simple and to the point
   - Original
   - Everlasting - able to reach across generations and be relevant for years to come
   - Easy to look at – produce positive emotions

   Your logo should be on everything involving your business: business cards, forms, letterheads, flyers, website, social media, etc.
2. **Develop a slogan.** This should be as few words as possible geared towards getting your customers and potential customers to recognize/think of that one thing for which you want your business to be known.
3. **Unify all aspects of your business with your brand and logo.** The same color scheme should be used on everything along with the same logo and slogan. All employees should know the slogan and understand its importance. The person who answers the phone should utilize these, and all email correspondence should have your slogan and brand contained within them.
4. **Be consistent.** Don't change your logo and slogan just to be changing it. Ideally, you will never need to make drastic changes. You want to remain recognizable and not confuse your customers with conflicting branding. Look at the Coca-Cola branding scheme. While the font may have changed slightly over the years, everyone worldwide knows the red background with white script is Coca-Cola, just as the world recognizes the Nike swoosh. If these two companies

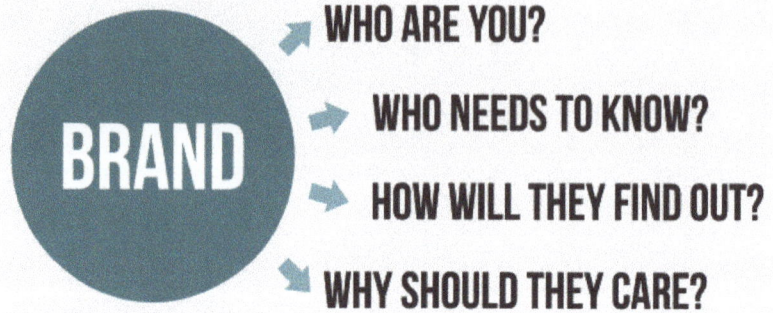

BRAND
→ WHO ARE YOU?
→ WHO NEEDS TO KNOW?
→ HOW WILL THEY FIND OUT?
→ WHY SHOULD THEY CARE?

had changed their branding strategies numerous times over the years, their logos would not be as recognizable as they are.
5. **Always have integrity.** Branding works both ways: if you don't deliver on promises and/or services, your brand will help those disgruntled customers remember you just as well as it will help your happy customers remember you.

Things to Avoid

When you begin a concerted effort to develop your brand, there are a few things to NOT do:
- Don't jump into creating your logo and slogan without a plan. If these aren't concise, you will end up with a (possibly many) logo that is difficult to remember or a confusing mess. If you don't have a plan outlining exactly what you are trying to accomplish, it will show.
- Don't make it more complicated than it has to be. Use clean, simple designs and phrases that are easy to remember, flow smoothly, and get your message across.
- Don't be vague. Don't develop a brand for the sake of having a brand. Slogans such as "best-selling" and "greatest" are a waste of time, as they have been over used in the past to the point that most people overlook them and move onto businesses with more attention-catching phrases.

Tips/Suggestions

The following are a few tips to help get you started in developing your brand:

- **Be Clear in your Brand Promise** – this lets your customers know what they can expect from you and why they need you.
- **Extricate your Business from Others** – draw attention to what makes you different from your competition.
- **Specialize** – point out your specialty
- **Connect with your target audience** – find common ground with the people whom you are trying to reach
- **Be Consistent** – everything that passes through your business and/or reaches your customers should consistently display your logo and slogan
- **Be Genuine** – do what you say you will do
- **Interact with your customers** – this can be done in a number of ways including social media, promotions, emails, surveys, etc.

Utilizing your Brand Strategy

Once you have your brand drilled down to a clear, concise logo and slogan, you must ensure your business and employees live up to its message. One bad experience a customer has with an employee can undo months or years of work building the brand. Be sure everyone in your business is on board with the brand and understands the importance of carrying out and demonstrating the message. This includes the way they interact with customers just as much as what they say to customers.

Next, you need to determine what channels you will use to get your branding message out to the community. This ties into your marketing strategy: they should work together. Where will you advertise to best reach your customers and potential customers?

There are many opportunities to build your online and offline communities including social media, trade shows, charity functions, etc. It is imperative to get your business name out there through marketing, and the manner in which you get it to people and ensure it is memorable is branding.

CHAPTER 22

Social Media

What Is Social Media?

Social media is the collective of online communications channels dedicated to community-based input, interaction, content-sharing and collaboration.

Perhaps best way to get a clearer understanding of it is to break it down into simpler terms.

The "social" part: refers to interacting with other people by sharing information with them and receiving information from them.

The "media" part: refers to an instrument of communication, like the internet (while TV, radio and newspapers are examples of more traditional forms of media).

Common Social Media Features

The following list shows features which are common to a social media site. If you are questioning whether a particular site may be classified as social or not, try looking for at least one of the below listed features.

User accounts: If a site allows visitors to create their own accounts where they can log in, then that is a good sign there is going to be social interaction. You can't really share information or interact with others online without doing it through a user account. Although not all sites that have a a user account are

social media, For instance, online retail/shopping, sites typically require a user account to conduct business, but most are not classified as being social media.

Profile pages: A profile page is often necessary to represent the individual. It usually has the option of including more information about the individual, such a profile photo, bio, website link, feed of recent posts, recommendations, recent activity and more.

Friends, followers, groups, hashtags and so on: Individuals use their accounts to connect with other users. They can also use them to subscribe to certain forms of information.

News feeds: When users connect with other users on social media, they are basically saying, "I want to get information from these people." That information is updated for them in real-time via their news feed.

Personalization: Social media sites usually provide users the flexibility to configure their user settings, customize their profiles to look a specific way, organize their friends or followers, manage the information they see in their news feeds and even give feedback on comments, articles, videos and more.

Notifications: Social media sites and apps often have the ability to notify users about specific information. Users have total control over these notifications and can choose to receive the types of notifications that they desire to obtain.

Information updating, saving or posting: If a site or an app allows users to post absolutely anything, with or without a user account, then it is probably social media. It could be a simple text-based message, a photo upload, a YouTube video, a link to an article or anything else.

Like buttons and comment sections: Two of the most common ways people interact on social media are via buttons that represent a 'like' or 'dislike' and comment sections.

Review, rating or voting systems: Besides liking and commenting, lots of social media sites and apps rely on the collective effort of the community to review, rate and vote on information that they know about or have used in the past. Shopping sites or movie review sites often use this social media feature.

What's the Difference Between Social Media and Social Networking?

The terms social media and social networking are often used interchangeably as if they mean the exact same thing. Although the difference is subtle, they are not the same. Social networking is a subcategory of social media.

The easiest way to understand the difference between social media and social networking is by thinking about the terms "media" and "networking" separately. Media refers to the information you are a sharing—whether it's a link to an article, a video, a pic, a PDF document, a simple status update or anything else.

Networking has to do with who your audience is and the relationships you have with them. Your network can include people like friends, relatives, colleagues, anyone from your past, current customers, mentors and even complete strangers. It refers to the group or demographic you are communicating with.

Traditional Media versus Social Media

Traditional media is the term used to encompass conventional forms of advertising **media** such as television, print, radio, direct mail and outdoor. While social **media** includes innovative and contemporary digital **media** like the Internet, email, mobile, blogging and social networking channels.

- **Traditional Media** is passive consumer participation. **Traditional Media** is one-way "one-to-many" communication. Marketers have used **traditional media** such as print, radio, TV, yellow pages and even outdoor ads to reach consumer markets for the last 50 to 100 years.
- **Social Media** is active consumer participation. **Social Media** is two-way "one-to-one" communication.

History of Social Media

In 1969 *CompuServe* was one of the first major commercial Internet providers in the US. They used the

Social Media	Traditional Media
Two-way conversation	One-way conversation
Open system	Closed system
Transparent	Opaque
One-on-one marketing	Mass marketing
About you	About ME
Brand and User-generated Content	Professional content
Authentic content	Polished content
FREE platform (usually)	Paid platform
Metric: Engagement	Metric: Reach/ frequency
Actors: Users/ Influencers	Actors / Celebrities
Community decision-making	Economic decision-making
Unstructured communication	Controlled communication
Real time creation	Pre-produced/ scheduled
Bottom-up strategy	Top-down strategy
Informal language	Formal language
Active involvement	Passive involvement
Deep Analytics	Poor analytics
Paid, Owned, Earned	Paid

technology known as dial-up to connect to the web. They were the dominant force during the eighties and part of the nineties until other companies eventually started catching up.

The first e-mail was sent in 1971 and was the prevailing way to communicate remotely for many years.

Although it is still used widely today, it is not as popular as in previous decades. Now, the popularity of text messaging, Skype, Facebook, Twitter and other sites communication websites give users an instant connection that characterizes the social media of our times.

In 1985, America Online (AOL) was founded and it became one of the most popular early providers of Internet connections.

Blogging started in 1997 and became a very popular social media medium. In the same year AOL Instant Message made its debut, allowing people to chat online. With demand new websites keep popping up as users continue to look for better ways to easily communicate with others.

The Different Types of Social Media

Websites and applications dedicated to forums, microblogging, social networking, social bookmarking, social curation, and wikis are among the different types of social media.

Following are some prominent examples of social media:

- *Facebook* is a popular free social networking website that allows registered users to create profiles, upload photos and video, send messages and keep in touch with friends, family and colleagues. According to statistics from the *Nielsen Group*, Internet users within the United States spend more time on *Facebook* than any other website.
- *Twitter* is a free microblogging service that allows registered members to broadcast short posts called tweets. Twitter members can broadcast tweets and follow other users' tweets by using multiple platforms and devices.
- *Google+* (pronounced *Google plus*) is *Google's* social networking project, designed to replicate the way people interact offline more closely than is the case in other social networking services.

The project's slogan is "Real-life sharing rethought for the web."

- *Wikipedia*, founded in January of 2001, is a free, open content online encyclopedia created through the collaborative effort of a community of users known as Wikipedians. Anyone registered on the site can create an article for publication; registration is not required to edit articles.

- *LinkedIn* is a social networking site designed specifically for the business community. The goal of the site is to allow registered members to establish and document networks of people they know and trust professionally.

- *Reddit* is a social news website and forum where stories are socially curated and promoted by site members. The site is composed of hundreds of sub-communities, known as "subreddits." Each subreddit has a specific topic such as technology, politics or music. *Reddit* site members, also

known as, "redditors," submit content which is then voted upon by other members. The goal is to send well-regarded stories to the top of the site's main thread page.

- *Pinterest* is a social curation website for sharing and categorizing images found online. *Pinterest* requires brief descriptions but the main focus of the site is visual. Clicking on an image will take you to the original source, so, for example, if you click on a picture of a pair of shoes, you might be taken to a site where you can purchase them. An image of blueberry pancakes might take you to the recipe; a picture of a whimsical birdhouse might take you to the instructions.

Brian Solis created the following social media chart, known as the conversation prism, to categorize social sites and services into various types of social media.

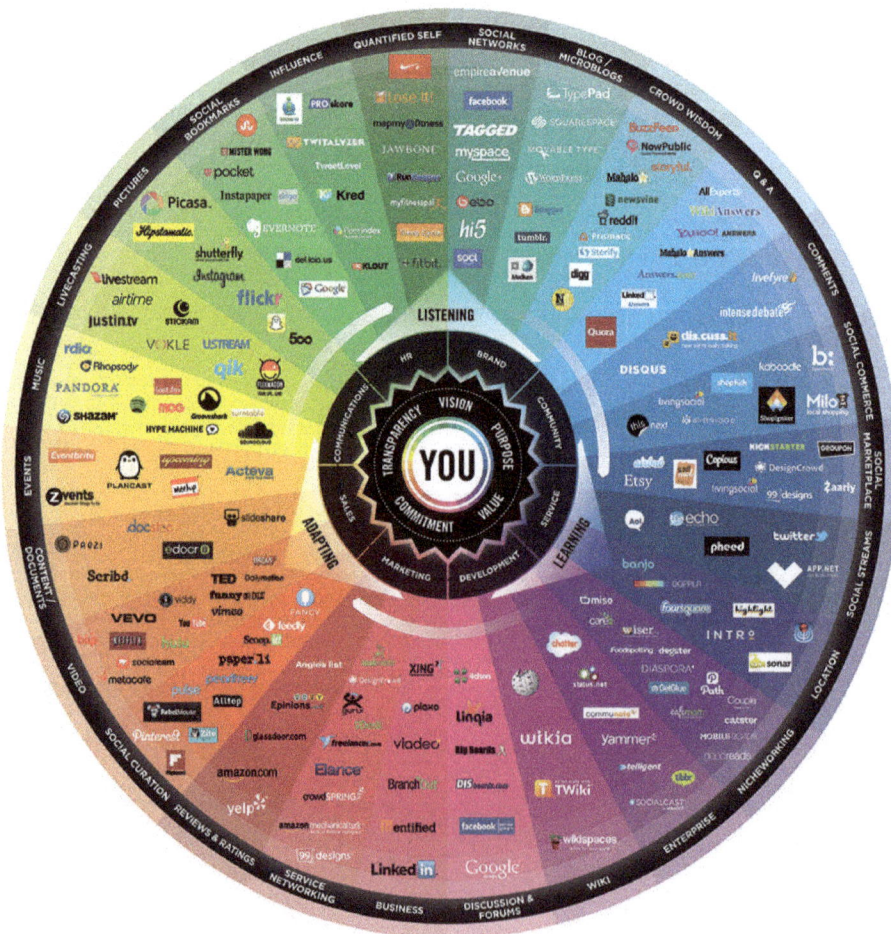

As social websites and applications proliferate social media is becoming an integral part of life for most people. Most traditional online media included social components, such as comment fields for users. In business, social media is used to market products, promote brands, connect to current customers and foster new business.

Social media analytics is the practice of gathering data from blogs and social media websites and analyzing that data to make business decisions. The most common use of social media analytics is to mine customer sentiment to support marketing and customer service activities.

Social media marketing (SMM) takes advantage of social networking to help a company increase brand exposure and broaden customer reach. The goal is usually to create content compelling enough that users will share it with their social networks.

One of the key components of SMM is social media optimization (SMO). Like search engine optimization (SEO), SMO is a strategy for drawing new and unique visitors to a website. SMO can be done two ways: by adding social media links to content such as RSS feeds and sharing buttons, or by promoting activity through social media via status updates, tweets, or blog posts.

Social CRM (customer relationship marketing) can be a very powerful business tool. For example, establishing a Facebook page allows people who like your brand and the way you conduct business to 'Like' your page, which creates a venue for communication, marketing and networking. Through social media sites, you can follow conversations about your brand for real-time market data and feedback.

From the customer's perspective, social media makes it easy to tell a company and everyone else about their experiences with that company -- whether those experiences are good or bad. The business can also respond very quickly to both positive and negative feedback, attend to customer problems and maintain, regain or rebuild customer confidence.

Enterprise social networking allows a company to connect individuals who share similar business interests or activities. Internally, social tools can help employees access information and resources they need to work together effectively and solve business problems. Externally, public social media platforms help an organization stay close to their customers and make it easier to conduct research that they can use to improve business processes and operations.

Social media is also often used for crowdsourcing. Customers can use social networking sites to offer ideas for future products or tweaks to current ones. In IT projects, crowdsourcing usually involves engaging and blending business and IT services from a mix of internal and external providers, sometimes with input from customers and/or the general public.

On the other hand, the integration of social media in the business world can also pose challenges. Social media policies are designed to set expectations for appropriate behavior and ensure that an employee's posts will not expose the company to legal problems or public embarrassment. Such policies include directives for when an employee should identify himself as a representative of the company on a social networking website, as well as rules for what types of information can be shared.

Advantages & Disadvantages of Social Networking

Social networking has shifted the way we communicate, do business, get our daily news and socialize. A site like *Facebook* can serve as an opportunistic launching pad for a new business owner, or it could be an inescapable source of negative peer pressure for a young teen. However, like most everything else in life, there are pros and cons social networking.

Below are some of the major advantages and disadvantages that most people who have spent time on the web can probably relate to in regards to social networking sites.

Advantages of Social Networking

- **Ability to connect with other people all over the world.** Perhaps the most obvious advantage of social networking is the convenience of connecting with people located anywhere. For instance, people can use *Facebook* to stay in touch with old friends who have relocated all over the country, get on *Google+* with relatives who live halfway around the world, or meet brand new people on *Twitter* from cities or regions they have never even heard of before.

- **Easy and instant communication.** Now that most people are 'connected' wherever they go, they don't have to rely on the home telephone answering machines or *U.S. Postal Service* to communicate. We can simply open up our laptops or pick up our smartphones and instantly communicate with anyone on any number of social media platforms or social messaging apps available.

- **Real-time news and information discovery.** Gone are the days of waiting around for the six o'clock news to come on, or for the delivery boy to bring the newspaper. If you want to know what is going on in the world, all you need to do is jump on social media. An added bonus is that we can customize our news and information discovery experiences by choosing the content we desire to follow.

- **Opportunities for businesses.** Business owners and other types of professional organizations have been able to connect with current customers, sell their products and expand their reach using social media. There are actually lots of entrepreneurs and businesses that thrive almost entirely on social sites and would not be profitable without it.

- **General fun and enjoyment.** For many, social networking is a fun. A lot of people turn to it when they are able to catch a break at work or just want to relax at home. Most people find it very satisfying to see comments and 'likes' show up on our own posts. People also find it to be a very convenient way to keep up with family and friends without having to ask them directly.

Disadvantages of Social Networking

- **Information over-load.** With so many people now on social media, tweeting links and posting selfies and sharing *YouTube* videos, life can become very noisy. Becoming overwhelmed by too many *Facebook* friends or too many *Instagram* photos to browse through isn't all that uncommon. Over time, we tend to rack up a lot of friends and followers, and that can lead to lots of bloated news feeds with too much content we're not all that interested in. All this 'noise' can easily become a distraction from one's goals and reduce an individual's production. This amounts to less time in prayer, reduced time exercising, and less face-time with meaningful people in our lives (i.e. spouse, children, immediate family, etc.).

- **Privacy issues.** With so much sharing going on, issues over privacy will always be a big concern. Whether it is a question of social sites owning your content after it is posted, becoming a target after sharing your geographical location online, or even getting in trouble at work after sharing too much online can open up all sorts of problems that sometimes can never be undone.

- **Social peer pressure and cyber bullying.** For people struggling to fit in with their peers -- particularly teens and young adults -- the pressure to do certain things or act a certain way can be even worse on social media than it is at school or any other offline setting. In some extreme cases, the overwhelming pressure to fit in with everyone posting on social media or becoming the target of a cyberbullying attack can lead to self-worth issues and even depression.

- **Online interaction substitution for offline interaction.** Since people are now connected all the time and you can pull up a friend's social profile with a click of your mouse or a tap of your smartphone, it is becoming a lot easier to use online interaction as a substitute for face-to-face interaction. Some people argue that social media actually promotes antisocial human behavior.

- **Distraction and procrastination.** How often do you see someone look at their phone? People get distracted by all the social apps and news and messages they receive, leading to all sorts of problems like distracted driving or the lack of gaining someone's full attention during a conversation. Browsing social media can also feed procrastination habits and become something people turn to in order to avoid certain tasks or responsibilities.

- **Sedentary lifestyle habits, sleep disruption, and electronic pollution.** Lastly, since social networking is all done on some sort of computer or mobile device, it can sometimes promote too much sitting down in one spot for too long. Likewise, staring into the artificial light from a computer or phone screen at night can negatively affect your ability to get a proper night's sleep. Furthermore, electrical pollution from wifi and other electric devices is harmful to human health. See 'Electro Pollution, WiFi

& EMF' at http://www.farmyourspace.com/recommended-resources/ for more information on this subject.

What Is Social Media Marketing?

Social media marketing refers to the process of gaining website traffic or attention through **social media** sites. **Social media marketing** programs usually center on efforts to create content that attracts attention and encourages readers to share it with their **social** networks.

Social media marketing is the use of social media platforms and websites to promote a product or service. Most of these social media platforms have their own built-in data analytics tools, which enable companies to track the progress, success, and engagement of ad campaigns. Companies address a range stakeholders through social media marketing including current and potential customers, current and potential employees, journalists, bloggers, and the general public. On a strategic level, social media marketing includes the management of the implementation of a marketing campaign, governance, setting the scope (e.g. more active or passive use) and the establishment of a firm's desired social media "culture" and "tone". To use social media effectively, companies need to allow customers and Internet users to post user-generated content (e.g., online comments, product reviews, etc.), also known as "earned media", rather than use marketer-prepared advertising copy. While social media marketing is often associated with companies, recently a range of not-for-profit organizations and government organizations started engaging in social media marketing to promote their programs, services, or agenda.

A corporate message spreads from user to user and presumably resonates because it appears to come from a trusted, third-party source, as opposed to the brand or company itself. Hence, this form of marketing is driven by word-of-mouth, meaning it results in earned media rather than paid media.

Social media has become a platform that is easily accessible to anyone with internet access. Increased communication for organizations fosters brand awareness and often, improved customer service. Additionally, social media serves as a relatively inexpensive platform for organizations to implement marketing campaigns.

There are two basic strategies for engaging the social media as marketing tools. There are the 'Passive Approach' and the 'Active Approach'. These strategies are described below:

Passive Approach

Social media can be a useful source of market information and a way to hear customer perspectives. Blogs, content communities, and forums are platforms where individuals share their reviews and recommendations of brands, products, and services. Businesses are able to tap and analyze the customer voices and feedback generated in social media for marketing purposes. As a result, social media is a relatively inexpensive source of market intelligence which can be used by marketers, companies, and other users to track and respond to consumer-identified problems and detect market opportunities.

Unlike traditional market research methods such as surveys, focus groups, and data mining which are time-consuming and costly, and which take weeks or even months to analyze, marketers can use social media to obtain 'live' or "real time" information about consumer behavior and viewpoints on a company's brand or products. This can be useful in the highly dynamic, competitive fast-paced and global marketplace.

Active Approach

Social media can be used not only as public relations and direct marketing tools but also as communication channels targeting very specific audiences with social media influencers and social media personalities and as effective customer engagement tools. Technologies predating social media, such as broadcast TV and newspapers can also provide advertisers with a fairly targeted audience, given that an ad placed during a sports game broadcast or in the sports section of a newspaper is likely to be read by sports fans. However, social media websites can target niche markets even more precisely. Using digital tools such as *Google Adsense*, advertisers can target their ads to very specific demographics, such as people who are interested in social entrepreneurship, political activism associated with a particular political party, or video gaming. *Google Adsense* does this by looking for keywords in social media user's online posts and comments. It is hard for TV stations or paper-based newspapers to provide ads that are this targeted.

Facebook and *LinkedIn* are major social media platforms where users can hyper-target their ads. This 'hypertargeting' not only uses public profile information but also information users submit but hide from others. There are many examples of companies initiating some form of online dialog with the public to foster relations with customers. For instance, business executives like Jonathan Swartz, President and CEO of Sun Microsystems, Steve Jobs CEO of Apple Computers, and McDonalds Vice President Bob Langert have regularly posted in their CEO blogs, encouraging customers to interact and freely express their feelings, ideas, suggestions, or remarks about their postings, the company or its products.

Using customer influencers can be a very efficient and cost-effective method to launch new products or services.

Engagement

In the context of the social web, engagement means that customers and stakeholders, such as consumer advocacy groups and groups that criticize companies (e.g., lobby groups or advocacy organizations) are active participants rather than passive viewers. Social media use in a business or political context allows all consumers or citizens to express and share an opinion about a company's products, services or business practices, or a government's actions. Each participating customer, prospect, or citizen who is participating online via social media actually become part of the marketing campaign, as others read their positive or negative comments or reviews. Getting consumers, potential consumers, or citizens to be engaged online is fundamental to successful social media marketing. With the advent of social media marketing, it has become increasingly important to gain customer interest in products and services, which can eventually be translated into buying behavior, votes, or a donation. New online marketing concepts of engagement and loyalty have emerged which aim to build community participation and brand reputation.

Engagement in social media for the purpose of a social media strategy is divided into two parts. The first is proactive, regular posting of new online content (digital photos, digital videos, text) and conversations, as well as the sharing of content and information from others via weblinks. The second part is reactive conversations with social media users responding to those who reach out to your social media profiles through commenting or messaging. Traditional media such as TV news shows are limited to one-way interaction with customers or 'push and tell' where only specific information is given to the customer with few or limited mechanisms to obtain customer feedback. Traditional media, such as paper newspapers, does give readers the option of sending a letter to the editor, but this is a relatively slow process, as the editorial board has to review the letter and decide if it is appropriate for publication. On the other hand, social media is participative and open, as participants are able to instantly share their views on brands, products, and services. Traditional media gave control of message to the marketer, whereas social media shifts the balance to the consumer (or citizen).

The Top 10 Benefits of Social Media Marketing

1. Increased Brand Recognition.
2. Improved brand loyalty.
3. More Opportunities to Convert.
4. Higher conversion rates.
5. Higher Brand Authority.
6. Increased Inbound Traffic.
7. Decreased Marketing Costs.
8. Better Search Engine Rankings.
9. Enrich relationship with customers.
10. Opportunity to gain valuable information about what your customers

Future Forecast of Social Media

1) Live video content will become more widely used.

Live video content is on the rise. According to *Social Media Examiner*, 43% more marketers plan to start using interactive video this in 2017. While there a lot of streaming sites and platforms available, both *Periscope* and *Facebook Live* are among the most popular in 2016; with statistics proving it. In its 2016 annual recap, *Periscope* noted that users watched 110 years of live video every day using the app. Live streaming on Facebook reached record-breaking numbers around the globe:

In addition to *Facebook Live* and *Periscope*, *Instagram* and *Twitter* launched their versions of live video streaming in November and December 2016, respectively.

As for the best ways to market, there are a lot of brands out there that are successfully using the live video strategy. This approach keeps their followers engaged with their brand by bringing an event they otherwise might not be able to attend in person.

Brands can also use live video for customer service by hosting Q&A sessions and product demonstrations. These videos drive engagement because hosts can ask for comments, questions, and feedback from the audience.

Brands can also stream multiple live videos in a series, providing more opportunities for engagement. *Facebook* statistics show that there is a 10 times higher rate of live videos using this approach.

2) Brands will use messaging apps much more in the future.

Many brands are using messaging apps to communicate one-on-one with customers, which is completely changing the way customer service gets done. These apps provide a faster and easier way for customers to get the assistance they need, rather than being placed on hold or waiting for a returned email. Deploying messaging for customer service is more scalable and cost-effective for businesses, and by providing a better experience for the customer, brands can solve their problems quickly and retain them more easily. For example, *Hyatt* uses *Facebook Messenger* for 24-hour customer service, where guests can make reservations, ask questions, and get recommendations for their trips.

Messaging apps such as *Facebook Messenger*, *WhatsApp*, and *WeChat* are used by 4 billion users worldwide. There are tremendous opportunities for brands to leverage their presence using message apps. With demand comes ingenuity. Already many new apps are coming on the market to address this need.

3) Social media ecommerce will become an much more powerful avenue for sales.

According to a survey from *Aimia*, 56% of consumers said they followed brands on social media to browse products for sale, and 31% of online shoppers say they are using social media specifically to look for new items to purchase.

Facebook, Instagram, Twitter, and *Pinterest* offer ways for users to purchase products from directly within their apps. *Snapchat* recently rolled out ecommerce features.

Companies will be increasingly changing the public's shopping habits using this social media strategy.

There will be an increase of sharing product pics and info with a "Buy Now" call-to-action; as well a gift ideas.

Product demonstration videos on social media will also become more popular. According to recent research from *Animoto*, four times as many customers prefer to watch a video about a new product before making a purchase.

4) Virtual reality will increasingly find its way into more marketing experiences.

Although virtual reality is still new to the marketing scene it is predicted by most tech marketing gurus in the industry to be widely used in the future. Companies are quickly recognizing the value that virtual reality brings with its offering of immersive, memorable experience unlike any other medium. A 360-degree virtual reality video has a tremendous impact on consumers by encouraging engagement and leaving a strong impression.

5) Ephemeral content will be used more by the public and by Marketers.

Ephemeral messaging is the mobile-to-mobile transmission of multimedia messages that automatically disappear from the recipient's screen after the message has been viewed. The word "ephemeral" describes something that only lasts for a short period of time.

Apps like *Snapchat*, *Periscope*, and *Meerkat* enable users to create content that is worthy of sharing, but not quite worthy of preserving; since most smartphone users are sharing more content intended for temporary consumption, while taking fewer permanent photos of things that matter more. With more people are using ephemeral content apps to share, they are a captive audience for what marketers.

6) Many brands will make the shift from Snapchat to Instagram for Stories.

Instagram recently introduced its *Stories* feature. After just two months it was reported that *Instagram Stories* were experiencing 100 million daily active viewers; which represents two thirds of *Snapchat's* total user base. This trend suggests that brands will start transitioning from *Snapchat* to *Instagram* for sharing *Stories* (photos and videos that disappear after 24 hours).

At 600 million users, *Instagram* offers a vastly bigger audience than *Snapchat* at 150 million users. Since *Facebook* owns *Instagram*, *Instagram* advertisers can target based on *Facebook* and *Instagram* insights, which means there is a bigger target audience on *Instagram* than *Twitter*.

Furthermore, *Instagram* lets users publish photos and videos in a permanent portfolio in addition to ephemeral *Stories*, so users can more easily share content with their friends. Celebrities and public figures are will also be adopting *Instagram* in greater numbers because of its slicker interface.

7) Mobile advertising will become much more competitive.

Marketers should expect greater investment in mobile advertising. Following is a summary of what is expected to transpire on the largest social networks:

Facebook is the behemoth when it comes to social media ad revenues, bringing in more than $7 billion 2016 -- 80% of which came from mobile ads. *Facebook's News Feed* algorithmic changes now prioritize content from friends and family first, so the 75% of brands on *Facebook* that pay to promote ads are expected to become more creative and design visual, engaging ads to get noticed first. In early 2017 *Facebook* began censoring conservative news, and promoting news from the 'left', along with hype of many fake news stories. Therefore, with more people becoming aware of *Facebook's* lack of integrity their reputation is suffering, which will hopefully open the doors to competitors who will operate without such political bias and scrupulous tactics.

Twitter's ad revenue is increasing, especially in the mobile format. They are expected to continue experimenting with visual content, such as sponsored hashtag icons and stickers, to provide a variety of ad options to users.

Snapchat and Instagram will be competing for a lot of attention and advertising revenue in the years ahead. *Snapchat* recently launched a new advertising API that makes it easier to buy ad space, in addition to a greater variety of video ads and sponsored filters. *Instagram* is doubling down on ecommerce with the introduction of *Shoppable Instagram*, a feature that lets users buy products directly by clicking on a CTA in the app.

According to a recent *Adweek* survey among millennial *Snapchat* and *Instagram* users about their experiences with ads, the results are roughly split -- with a few noteworthy distinctions. While a greater percentage thought *Instagram* ads were more memorable than those on *Snapchat*, more millennials liked *Snapchat* ads than *Instagram* ads:

In summary, companies should experiment with ads on different platforms to see which perform better among their audience and take advantage of the new technical features that are on the horizon.

How to Make Social Media work for You

Leveraging the power of content and social media marketing can help increase your customer base in a dramatic way. To get the most benefit out of social media marketing it is vital that you understand the industry fundamentals. From maximizing quality to increasing your online entry points, abiding by the following 'best practices' principles will help build a foundation that will serve your company, customers, brand, and profit margin.

1. **Select the best social media platform(s) for your business** – Trying to reach customers on every platform is typically not the best approach for most companies. Some platform will not be a good 'fit'. For example, a tech startup is more likely to find new customers on sites like *Twitter* than sites like *Pinterest*, which more to people that are more into crafts and domestic type interests. Knowing your audience is the first step to success with social media.

2. **Start with the end in mind** – Jumping blind into social media is just as ill-advised as flying blind into any other aspect of your business. Take the time to have a clear plan in mind before you start posting to avoid losing direction and dissuading people from following your business.

3. **Quality over quantity** – It is better to have 1,000 online connections in your target market who read, share and talk about your content within their own circles than 10,000 contacts who are not a good 'fit' for your company or agenda.

4. **Ask questions to boost involvement** – Social media is built around a sense of community, and the best way to succeed is to *create a community* and interact with it. Posting questions in your content will increase comments and interest. People like to get involved, help, and voice their opinion. Building a community is essential to making sure your posts are seen, since

the algorithms on social media sites respond to interaction on posts.

5. **The importance of listening** – Success with social media and content marketing often requires more listening and less talking. Read your target audience's online content and join discussions to learn what is important to them. Doing so will empower you to be able to create content and generate conversations that add value rather than clutter to their lives.

6. **The importance of patience** – Social media marketing success doesn't happen overnight. While it is possible to have your message go viral, it is far more likely to take time and much effort to achieve your goals. Therefore, it is important to be committed to the long-term journey to achieve your desired results.

7. **The law of compounding** – If you publish amazing, quality content and work to build your online audience of quality followers, they will share it with their own audiences on Twitter, Facebook, LinkedIn, their own blogs and more. This sharing and discussing of your content opens new entry points for search engines like Google to find your products, services, and company in keyword searches. Those entry points can lead to exponential growth in many potential ways for people to find you online.

8. **The importance to connections and influence** – Spend time finding the online influencers in your market who have quality audiences and are likely to be interested in your products, services and business. Connect with those people and work to build relationships with them. If you get on their radar as an authoritative, interesting source of useful information, there is a higher probability they will share your content with their own followers, which would put you and your business in front of even more people.

9. **The law of value** – If you spend all your time on the social Web directly promoting your products and services, people will tune you out. You must add value to the conversation. Therefore, it is important that you focus less on conversions and more on creating content that adds to people's lives, as well as developing relationships with online influencers. In time, those people will become a powerful catalyst for word-of-mouth marketing for your business.

10. **The law of acknowledgment** –Building relationships is one of the most important parts of social media marketing success, so always acknowledge every person who reaches out to you.

11. **The importance of accessibility** – Don't publish your content and then disappear. Be available to your audience. Consistently publishing value adding content and participating in associated conversations are keys to success. Followers online can be fickle and they won't hesitate to tune you out, replace you, or forget about you if you disappear from being online for an unusual amount of time.

12. **The importance of reciprocity** – You can't expect others to share your content and talk about you if you don't do the same for them. Therefore, a portion of the time you spend on social media should be focused on sharing and talking about content published by others and/or helping to promote others in some way.

13. **Measure your results and act accordingly** – Having good analytics of all of your posts is one thing, but actually using the data you obtain to improve your business's social media strategy is another. Businesses that have a good social media presence use the analytics as a tool to learn, continually grow and adapt their market strategy to make sure they are always improving and staying abreast of market and social media changes.

Enticing Customers with Incentives

Whatever the reason, incentives such as discounts, coupons, contest, etc. entice people to engage. These engagements provide further marketing opportunities. Following are some ways in which you can better take advantage of this phenomenon.

1. **Appoint a qualified trustworthy individual to manage your social media accounts.** Although the younger generation has grown up with social media and tends to have a very good understanding of it, it is important to be very carefully who you have communicating with the public on behalf of your company. Make sure that whoever is managing your social media accounts---whether it is someone within your family, company or outsourced----understands the best way to convey your messages, promote your company, and achieve your goals.

2. **Implement a social media plan.** Take the time to craft well-planned messages that will showcase your company well. Getting into a routine of planning a week's worth of posts will help keep you on the right track of posting valuable content and take the stress out of trying to come up with a post just for the sake of posting. Services, such as HubSpot or HootSuite, can be good aids for scheduling future posts.

3. **Drive people to your website.** Your social media accounts should entice and attract people your website. You can attract visitors to your website by providing thoughtful, quality content that adds value and that speaks to your buyers needs. These types of social media messages will encourage them to visit your website to learn more, and seek out your products or services.

4. **Promote, but don't overdo it. A good rule of thumb is to f**ollow the 60/30/10 rule: Post 60% relevant, third-party industry information; 30% your own blogs to demonstrate your industry thought leadership; and 10% direct content offers. Any more than 10% typically comes across as being too 'sales-y' and will cause a drop in followers.

5. **Engage with your customers and prospects.** Social media is about being social. Show the people who reach out to you that you value their input by giving them some of your time and responding to them, even if you say nothing more than "I appreciate your comment, (first name)!" If you stay on top of it, social media engagement is one of the easiest, best, and least expensive ways to leave a customers and prospects feeling positive about your company, products or services.

CHAPTER 23

Motivating Employees

Motivated employees are more productive which, in turn, makes a business more successful. However, motivating employees can be a challenge for managers and supervisors, so how do you do it? The first step is to hire individuals who are motivated; however, you don't always have this luxury, and even when you do, you must keep them motivated by creating a work culture that motivates.

Incentives

The most commonly used tool in motivating employees is the use of incentives. Incentives are used to encourage employees to do their best work, knowing that when a project is done well they will receive these tokens of appreciation. The following are a few examples of incentives.

- Snack day at the office
- Catered meal to the office
- Day off
- Gift cards
- Movie tickets
- Tickets to sporting events
- Tickets to liberal arts activities
- Cash

While these are all good things and can (and probably should) be used as motivation boosters, they are short-term ideas and do not encourage long-term motivation. **People want to be treated as if they matter**, and when they are not treated as such, they will shut down and not be motivated to do anything.

Creating a Work Culture that Motivates

The golden rule of treat others as you want to be treated definitely applies here and captures the essence of the best way to create a motivating work culture.

You give your opinion of your employees through your words, body language and facial expressions when you interact with them. To ensure you are doing your part to motivate employees, you should smile more, listen to your employees, earn their trust, and show them you genuinely care about them.

There are also specific actions you can take in the work place to help keep employees motivated:

- **Lead by example** – show up and be on time; demonstrate a positive attitude
- **Make decisions** – don't be wishy-washy; make a decision and stick to it (unless presented with

valid reasons not to – then communicate why you changed your decision)

- **Plan and execute projects** – be organized and communicate the plan to employees
- **Provide proper progressive discipline** – don't tolerate someone not doing their share of the work (This is possibly the number one reason employees get "turned off" of work.)
- **Delegate tasks and projects** – give employees opportunities to lead; this helps them feel empowered
- **Listen** – when an employee comes to you with a concern/suggestion truly listen and address it; don't appear to listen and then forget it
- **Inspire** – show employees the importance of their work: final product, client testimonials, etc.
- **Give clear instructions** – employees get very frustrated when they have no clear direction
- **Praise and recognize** – a simple "good job" goes a long way with employees
- **Spend time with employees** – spending one-on-one time with employees shows them they are valuable
- **Genuinely care** – take an interest in your employees both on the job and in their personal lives
- **Take an interest in employees' careers** – help employees prepare for their future and possibly move up their career ladder by providing opportunities for training; cross training employees is also an excellent way to help employees prepare for other jobs within the department/business
- **Take an interest in work-life balance** – encourage employees to balance their work and personal lives in a healthy manner; use days off as an incentive or recognition for a job well done
- **Give employees responsibilities** – this goes along with delegating tasks and empowering employees
- **Encourage self-care** – allow for time to exercise during the day

While there are many things you can do to encourage your employees to be motivated (as evidenced above), there are also a few things managers/supervisors/bosses do that immediately crush motivation in employees. Oftentimes people do these without realizing it or understanding the effects they have on others. The following is a list of these absolute motivation crushers.

Do NOTS

- **Micromanage** – When employees are micromanaged, they feel they are not trusted to do the job. Why think on their own, if someone is going to be looking over their shoulder every step of the way?
- **Lose your temper** – People never like to be on the receiving end of a temper tantrum; yelling, belittling and berating employees ensure you will lose their respect.
- **Be negative** – Being constantly negative and using negative comments toward employees quickly creates a negative workplace and ensures loss of respect.
- **Demonstrate lack of respect for employees** – Being late for meetings, not returning employee messages/phone calls, and not addressing/following through on employee concerns or suggestions shows them you have no respect for them. Why should they respect you or give their all on the job, if you don't respect them?
- **Not give credit where credit is due** – If you take credit for every good thing that comes out of your business/department, you will quickly lose respect and effort from the employees who actually deserve it.
- **Cave under pressure** – If you don't support your employees and back them up to others, you will quickly lose their respect.
- **Be Sparse with praise** – If employees never hear "good work" or something similar, they will lose

confidence in themselves and you, and they will lose the desire to do good work.

- **Not follow through** – Say what you mean, and mean what you say, then prove it. If you don't, your employees will not believe in you or the work. Therefore, they will not be motivated.
- **Not be trustworthy** – This goes along with saying what you mean and meaning what you say, but it also addresses keeping confidences of employees. Your employees should be able to talk to you about issues they may have at work (and possibly elsewhere), and trust that you will not share those with your co-workers.

Rules to a Motivating Work Culture

The most important, overarching ingredient to motivating employees is developing relationships with them. Get to know your employees. Are they married? Do they have children? What are their career goals? What do they enjoy doing outside of work?

Some of this can be accomplished during one-on-one meetings with the employee such as during performance reviews or during progress checks on a project. However, most of this relationship building comes through daily interaction with the employee through simple inquiries such as "How was your weekend;" "How are the kids;" etc. You are not there to be their friend, but it is important to understand them and know what motivates them. This can be instrumental to the success of the business

Once you have developed this kind of relationship with your employees, it is natural to mentor, appreciate, recognize, encourage, coach, and care for them. Positive interaction with employees encourages motivation and brings out the best in them, allowing them to excel in the workplace.

Another advantage to getting tp know your employees is that you understand what motivates them, because you've learned about their families, their interests, their goals in life. And employees with

a strong connection to their managers are more likely to work longer hours and be loyal to the company.

However, in order to be a good manager, you must be careful to distinctly define the boundaries between yourself and your staff. Following are some points to remember:

- **Clarify the relationship.** To maintain the respect of your employees while being friends with them, you must be direct about the nature of your business relationship. This means being clear about what the goals are, how your employees are to help you accomplish them, and what they can expect from you. By communicating these things clearly, you curtail the risk that an employee can misinterpret your friendship and behave in an unprofessional manner.
- **Be social, to a degree.** Socialize with all employees as equally as possible, and don't show favoritism. Consuming alcohol with employees or being the last on at the party are invitations to problems. Also, keep socializing at the office to a minimum. You want to ensure that you are respected as well as liked.
- **Don't play favorites.** One of the worst mistakes you can make is to favor certain employees in the workplace. Your other staff members will quickly learn to distrust you, and productivity will suffer.
- **Keep confidential information confidential.** No matter how close you are to pals in the office, you have to resist the temptation to give them the inside scoop. Confidential work information like salaries, hiring and firing decisions, and quarterly earnings must *never* be shared with others or you will lose credibility.
- **Confront poor performance.** Sometimes an employee who may be considered a friend my not perform as expected. For the sake of your company, you need to consider how this person's behavior is impacting employee morale, work schedules, customer relations, time spent fixing mistakes, and most importantly, the bottom

line. In such situations, you must be this person's employer first and friend second. If you can help this staff member return to being a productive member of your team, then do so. If not, you need to let the person go before more damage is done.

- **Don't fake it.** Asking staff about their personal lives, such as their weekend plans, their families, or their children, is a good way build loyalty. However, such efforts can backfire if it is viewed as being insincere. Its okay to ask occasional questions of staff, but don't make a big production out of it. Getting to know people takes time.

- **Make sure people understand their job responsibilities in detail.** This helps employees know their role and when they may be stepping outside those boundaries. It also helps you measure their job performance. Providing employees with a written job description is a good way to clearly define their roles and responsibilities.

- **Let your employees know what type of work behavior is unacceptable**. In addition to our organization's policies and procedures, you may have your own workplace rules and conduct expectations. Examples of unacceptable behavior may include the use of cell phones during work hours, offensive language, gossiping about other employees and/or other topics of conversation. Make sure you communicate the consequences for engaging in unacceptable behavior. The best thing you can do is to stop the behavior immediately and let your know employee know what they did was wrong and to not do it again

- **Lead by example.** Try to set the tone by avoiding conversations about non-work or personal matters when employees are "on the clock". Remember that the rules you set for your employees should apply to you as well. Leading by example is one of the best ways to reinforce desired behavior within your care team.

- **Set boundaries.** Define boundaries and guidelines for employees during an all-staff meeting.

In the alternative, convene small groups of employees to facilitate better discussion about boundaries, guidelines and penalties for employees who ignore them.

Dangers of Fraternization

Caring about and getting to know employees will produce positive results. However, boundaries and rules of engagement need to be established or fraternization and its unfortunate consequences will undoubtedly occur.

Fraternization in the workplace encompasses relationships that go beyond the normal scope of employee interactions. The problems typically creep in when the fraternization occurs between a supervisor and subordinate, whether the relationship is romantic in nature or simply a strong friendship. Realizing the dangers of fraternization helps you maintain a personal policy prohibiting the type of interactions that go too far. Following are the most common problems that can occur as a result of fraternization.

Favoritism – When an employee and her supervisor become close outside of work, favoritism is a possibility, whether intentional or not. People naturally want to protect and support those they are close with. In such cases, the manager then tends to allow more leniency or privileges to the subordinate. Promotions, rebukes, and discipline become more difficult, and other employees can become resentful.

Tension – Personal relationships often go through difficult periods. Any tension in the personal relationship is likely to carry over to the workplace. The two people involved may find it difficult to work together or even argue at work. If the relationship dissolves, further tension is likely to be felt at the office.. These negative dynamics can radiate throughout the workplace causing many problems and a decrease in production.

Legalities – Fraternization lead to many problems in the workplace. It also opens up the opportunity for legal problems if the relationship with the employee(s) in question becomes tarnished or if other employees feel slighted. In such cases the manager and company may find themselves faced with a law suit.

Conclusion

Human beings are social creatures who thrive on companionship and recognition. Consideration and respect in the workplace increases creative output and induces employees to be more productive and motivated. Every person is motivated by something. Find what motivates each of your employees and work with it to ensure a positive work place. Having motivated employees is key to a successful business; however, the bottom line is employees (people, in general) have to motivate themselves. They have to want to do the work to the best of their ability. Your job is to create a work culture that encourages this.

CHAPTER 24

Dressing for Success

What is Dressing for Success?

Dressing for Success is portraying the image you want peers, bosses and potential future bosses to see. You want to come across as capable and confident. The clothes you wear can go a long ways towards this goal and often times help you appear capable and confident even during times when you may not actually feel it.

You should dress in a way that represents who you are and who you want to be. While the education, experience, and qualifications probably landed you an interview, how you present yourself in the interview will most likely determine whether or not you get the job. If you already have a job, you should be dressing for the job you want next.

Why is Dressing for Success Important?

Most of us are familiar with the old saying there is no second chance to make a good first impression. This is why dressing for success is so important. You always want to put your best foot forward; you never know when an opportunity may arise from a chance meeting.

Dressing up tends to make people more productive, creative, and successful. Wearing nicer clothes typically raises one's confidence and definitely affects how others perceive you. How you dress tells others whether or not you are serious about your work, whether or not you are professional in your work, and whether or not you have the confidence and power to be a good leader.

WE WANT YOU!

How to Dress for Success

Interviews are the optimal place to make a good first impression. When interviewing for a professional position with a traditional company, dress in your best business attire. When interviewing with more casual businesses, wear business casual attire. It is ideal to visit the company before your interview to see how current employees dress.

Business attire = solid color, conservative suits, modest shoes, light pantyhose and limited jewelry for women; solid color, conservative suits, professional shoes with matching socks and conservative ties for men

Business casual = slacks (khaki and corduroy are okay) pants or skirts, sweaters, cardigans, knit shirts, and conservative shoes for women; slacks/pants (khaki, cotton, gabardine), long-sleeved button-down shirts (tie optional), knit shirts with a collar, sweaters, leather shoes, and a belt for men. *Note that Business Casual can mean different things with different companies; therefore, if you are unable to visit the company before your interview ask what is appropriate.

Tips for Dressing for Success

The following tips will help you feel more comfortable and confident in your interview and help you be more productive and successful in your job.

1. First and foremost be clean.
2. Combed/styled hair
3. Clean fingernails, if wearing fingernail polish be sure it's not chipped
4. Light makeup, if any
5. Avoid bloodshot eyes by getting a good night's sleep the night before.
6. Be sure your clothes are clean and pressed.
7. No stains, smudges, wrinkles.
8. No holes/runs in socks or hose.
9. Be sure your clothes fit correctly. Baggy clothes project a sloppy, lazy image. Tight-fitting and/or skin-revealing clothes give the impression you are not very professional.
10. Don't forget about your shoes. These should be cleaned and polished, if applicable. No dirt, scratches, stains.
11. Best colors to portray professionalism are blue, black and darker shades.
12. Solids and pinstripes are generally considered more professional than plaids.
13. Minimal jewelry. Wedding bands and one additional ring – not a ring on every finger. Conservative earrings as opposed to large hoops. One bracelet (possibly three if in a classy stack) as opposed to ten up and down your arm.
14. Cover your tattoos as much as possible. Tattoos are becoming more acceptable in some businesses, but it is best to cover them up, until you secure the job and see how the work culture feels about them.
15. Pockets should be empty to avoid bulges in your professional look
16. Be sure your cell phone is turned off.

Absolute DO NOTS:

The following are inappropriate for interviews and the work place under almost all circumstances:

- Crocs
- Uggs
- Stiletto heels
- Flip-flops
- Hats (unless required for the job or part of a uniform)
- Heavy make-up
- Excessive jewelry
- Radical jewelry
- Sunglasses (unless working outdoors)
- Sweatpants
- Yoga pants
- Leggings
- Shorts
- Short skirts
- Sweatshirts
- T-shirts with messaging
- Sports jerseys
- Tank tops
- Sequins (minimal decoration is okay)
- Glitter
- Animal prints
- Gum chewing

Pitfalls to Avoid

Be aware of Casual Fridays. Even though your company allows casual dress on Fridays, be sure you will feel comfortable in your clothing choice, if you have to meet a client unexpectedly or attend an unscheduled meeting. Also, be careful to not get too casual.

Don't be a Groupie. If others in your office wear jeans, tennis shoes, flip-flops or any other unprofessional attire don't join them, even when you believe they are not suffering any repercussions. Set yourself above and apart.

Don't Be Recognized for the Wrong Reasons. Wearing tight, clingy, revealing clothing may get you recognized, but not in the professional way you are seeking.

Additions to Dressing for Success

Appearance is second only to communication skills in qualities most associated with professionalism. While donning your professional attire, be sure to have steady eye contact in discussions, stand erect with your shoulders back to portray confidence, and practice firm handshakes. Remember to practice appropriate business manners such as not interrupting others, not walking into offices unannounced, not gossiping, always being on time, and always being prepared. Dressing for success involves a number of elements in addition to clothes. Your professional attire along with appropriate body language and business etiquette completes the positive image you're seeking to be successful.

Religious Expression in the Workplace

While it is important to be able to express your religious beliefs, the workplace is not the best place for this. There are laws giving you certain rights to wear religious clothing and items at work, providing you are not infringing on other people's rights or putting others in danger. Employers have the right to enforce dress codes, as long as the dress code does not discriminate against a specific group. If you are considering dressing to express your religious beliefs, there are several things to consider before doing so.

First, religious clothing is not recommended in job interviews. Unless you are applying at a religious organization, it is best to keep your religion to yourself, at least until you become familiar with the work culture you are entering. Second, expressing religious beliefs through your clothing may hurt work relationships and business in general by alienating or even driving away potential clients or customers. You are a representative of the company for which you work. Third, keep in mind you may be opening yourself up to conflict and personal issues, if your peers and/or bosses do not share your religious beliefs.

Be sure you are willing to face the possible consequences before opting to wear religious clothing in the work place. It may keep you from getting a job, decrease your productivity on the job, disrupt the work culture, or even cause you emotional and/or physical duress.

Conclusion

Dressing better makes you feel better and gives you more self-confidence and self-respect. It effects how others see you and gives them more confidence in you. Professional dress demonstrates your attention to detail and keeps you productive. Some may say dressing more comfortably will make you more productive; however, casual dress tends to put you in a comfort zone where you become lazy and much less productive.

If you are unsure of the dress code for an event or job, over dress rather than going casual. Be sure to maintain a clean, well-groomed appearance. A work wardrobe doesn't have to be expensive. Look for classic, professional pieces. You want good quality clothing, but no matter how much you spend, stains, tears, and wrinkles will immediately negate any amount of money you may have spent.

Be mindful of a balance between individual freedom and success. Many people believe they should have the right to express their individuality in their grooming and dress; however, as an employee, you represent the company for which you work. You are a walking image of the company, and as such should create a positive impression. Dress for success if you desire to be successful.

The majority of successful people are well-dressed. Your clothing is a form of communication and strongly influences how people perceive you. You want to put the entire package together to ensure your success.

CHAPTER 25

Risk Management

Projects often get started in the right direction but then get off track. For example, project managers will spend time with their teams to develop a clear scope and detailed plan. Then something happens; something unexpected—a major disaster strikes. The project manager and team move quickly into their reactive mode – they manage this risk based on their experiences and best judgment but they have no opportunity to test it out and they hope that it'll be okay, but they do not know for sure. This is not risk management – it is management by crisis

What Is Risk Management?

Risk Management is the process of identifying, analyzing and responding to risk factors throughout the life of a project and in the best interests of its objectives. Proper risk management implies control of possible future events and is proactive rather than reactive.

Risk Management Systems

Risk Management Systems are designed to do more than just identify the risk. The system must also be able to quantify the risk and predict the impact of the risk on the project. The outcome is therefore a risk that is either acceptable or unacceptable. The acceptance or non-acceptance of a risk is usually dependent on the project manager's tolerance level for risk.

If risk management is set up as a continuous, disciplined process of problem identification and resolution, then the system will easily supplement other systems. This includes; organization, planning and budgeting, and cost control. Surprises will be diminished because emphasis will now be on proactive rather than reactive management.

Risk Management is a Continuous Process

Once the Project Team identifies all of the possible risks that might jeopardize the success of the project, they must choose those which are the most likely to occur. They would base their judgment upon past experience regarding the likelihood of occurrence, gut feel, lessons learned, historical data, etc.

Early in the project there is more at risk then as the project moves towards its close. Risk management should therefore be done early on in the life cycle of the project as well as on an on-going basis.

The significance is that opportunity and risk generally remain relatively high during project planning (beginning of the project life cycle) but because of the relatively low level of investment to this point, the amount at stake remains low. In contrast, during project execution, risk progressively falls to lower levels as remaining unknowns are translated into knowns. At the same time, the amount at stake steadily rises as the necessary resources are progressively invested to complete the project.

The critical point is that Risk Management is a continuous process and as such must not only be done at the very beginning of the project, but continuously throughout the life of the project. For example, if a project's total duration was estimated at 3 months, a risk assessment should be done at least at the end of month 1 and month 2. At each stage of the project's life, new risks will be identified, quantified and managed.

Risk Response

Risk Response generally includes:

- **Avoidance** – eliminating a specific threat, usually by eliminating the cause.
- **Mitigation** – reducing the expected monetary value of a risk event by reducing the probability of occurrence.
- **Acceptance** – accepting the consequences of the risk. This is often accomplished by developing a contingency plan to execute should the risk event occur.

In developing Contingency Plans, the Project Team engages in a problem solving process. The end result will be a plan that can be put in place on a moment's notice.

What a Project Team would want to achieve is an ability to deal with blockages and barriers to their successful completion of the project on time and/or on budget. Contingency plans will help to ensure that they can quickly deal with most problems as they arise. Once developed, they can just pull out the contingency plan and put it into place.

Why do Risk Management?

The purpose of risk management is to:

- Identify possible risks.
- Reduce or allocate risks.
- Provide a rational basis for better decision making in regards to all risks.
- Plan.

Assessing and managing risks is the best weapon you have against project catastrophes. By evaluating your plan for potential problems and developing strategies to address them, you'll improve your chances of a successful, if not perfect, project.

Additionally, continuous risk management will:

- Ensure that high priority risks are aggressively managed and that all risks are cost-effectively managed throughout the project.
- Provide management at all levels with the information required to make informed decisions on issues critical to project success.

If you don't actively attack risks, they will actively attack you!!

How to Do Risk Management

First we need to look at the various sources of risks. There are many sources and this list is not meant to be inclusive, but rather, a guide for the initial brainstorming of all risks. By referencing this list, it helps the team determine all possible sources of risk. Various sources of risk include:

- **Project Management**
  ◊ Top management not recognizing this activity as a project
  ◊ Too many projects going on at one time

- Impossible schedule commitments
- No functional input into the planning phase
- No one person responsible for the total project
- Poor control of design changes
- Problems with team members.
- Poor control of customer changes
- Poor understanding of the project manager's job
- Wrong person assigned as project manager
- No integrated planning and control
- Organization's resources are overcommitted
- Unrealistic planning and scheduling
- No project cost accounting ability
- Conflicting project priorities
- Poorly organized project office

- **External**
  ◊ Unpredictable
    - Unforeseen regulatory requirements
    - Natural disasters
    - Vandalism, sabotage or unpredicted side effects
  ◊ Predictable
    - Market or operational risk
    - Social
    - Environmental
    - Inflation
    - Currency rate fluctuations
    - Media
  ◊ Technical
    - Technology changes
    - Risks stemming from design process
  ◊ Legal
    - Violating trade marks and licenses
    - Sued for breach of contract
    - Labour or workplace problem
    - Litigation due to tort law
    - Legislation

The Risk Analysis Process

The Risk Analysis Process is essentially a quality problem solving process. Quality and assessment tools are used to determine and prioritize risks for assessment and resolution. The risk analysis process is as follows:

- **Identify the Risk**
  ¤ This step is brainstorming. Reviewing the lists of possible risk sources as well as the project team's experiences and knowledge, all potential risks are identified.
  ¤ Using an assessment instrument, risks are then categorized and prioritized. The number of risks identified usually exceeds the time capacity of the project team to analyze and develop contingencies. The process of prioritization helps them to manage those risks that have both a high impact and a high probability of occurrence.

- **Assess the Risk**
  ¤ Traditional problem solving often moves from problem identification to problem solution. However, before trying to determine how best to manage risks, the project team must identify the root causes of the identified risks.
  ¤ The project team asks questions including:
    - What would cause this risk?
    - How will this risk impact the project?

- **Develop Responses to the Risk**
  ¤ Now the project team is ready to begin the process of assessing possible remedies to manage the risk or possibly, prevent the risk from occurring. Questions the team will ask include:
    - What can be done to reduce the likelihood of this risk?
    - What can be done to manage the risk, should it occur?

- **Develop a Contingency Plan or Preventative Measures for the Risk**
  - ¤ The project team will convert into tasks, those ideas that were identified to reduce or eliminate risk likelihood.
  - ¤ Those tasks identified to manage the risk, should it occur, are developed into short contingency plans that can be put aside. Should the risk occur, they can be brought forward and quickly put into action, thereby reducing the need to manage the risk by crisis.

10 Rules for Managing Project Risk

The Risk Management Process is intended to reduce management by crisis. While there may always be some things that will occur that are unanticipated, most of these, through sound risk management, can be managed, rather than reacted to. Essentially, the Risk Management Process is a quality problem-solving process. Quality and assessment tools are used to determine and prioritize risks for assessment.

1. Identify the risks early on in your project

- Review the lists of possible risk sources as well as the project team's experiences and knowledge.
- Brainstorm all potential risks.
- Brainstorm all missed opportunities if project is not completed.
- Make clear who is responsible for what risk.

2. Communicate about risks

- Pay attention to risk communication and solicit input at team meetings to ensure that risk management is perceived as important for the project.
- Focus your communication efforts with the project sponsor or principal on the big risks and make sure you don't surprise the boss or the customer.
- Also, make sure that the sponsor makes decisions on the top risks, because some of them usually exceed the mandate of the project manager.

3. Consider opportunities as well as threats

- While risks often have a negative connotation of being harmful to projects, there are also "opportunities" or positive risks that may be highly beneficial to your project and organization. Make sure you create time to deal with the opportunities in your project. Chances are your team will identify a couple of opportunities with a high pay-off that may not require a big investment in time or resources. These will make your project faster, better and more profitable.

4. Prioritize the risks

- Some risks have a higher impact and probability than others. Therefore, spend time on the risks that cause the biggest losses and gains. To do so, create or use an evaluation instrument to categorize and prioritize risks.
- The number of risks identified usually exceeds the time capacity of the project team to analyze and develop contingencies. The process of prioritization helps the project team to manage those risks that have both a high impact and a high probability of occurrence.

5. Assess the risks

- Traditional problem solving often moves from problem identification to problem solution. However, before trying to determine how best to manage risks, the project team must identify the root causes of the identified risks.

- Risk occurs at different levels. If you want to understand a risk at an individual level, think about the effect that it has and the causes that can make it happen. The project team will want to ask questions including:
  - ◊ What would cause each risk?
  - ◊ How will each risk impact the project? (i.e., costs? lead time? product quality? total project?)
- The information you gather in a risk analysis will provide valuable insights in your project and the necessary input to find effective responses to optimize the risks.

6. Develop responses to the risks

Completing a risk response plan adds value to your project because you prevent a threat occurring or minimize the negative effects. To complete an assessment of each risk you will need to identify:

- What can be done to reduce the likelihood of each risk?
- What can be done to manage each risk, should it occur?
- What can be done to ensure opportunities are not missed?

7. Develop the preventative measure tasks for each risk

- It's time to think about how to prevent a risk from occurring or reducing the likelihood for it to occur. To do this, convert into tasks, those ideas that were identified to reduce or eliminate risk likelihood.

8. Develop the contingency plan for each risk

- Should a risk occur, it's important to have a contingency plan ready. Therefore, should the risk occur, these plans can be quickly put into action,

thereby reducing the need to manage the risk by crisis.

9. Register project risks

- Maintaining a risk log enables you to view progress and make sure that you won't forget a risk or two. It's also a communication tool to inform both your team members, as well as stakeholders, what is going on.
- If you record project risks and the effective responses you have implemented, you create a track record that no one can deny, even if a risk happens that derails the project.

10. Track risks and associated tasks

- Tracking tasks is a day-to-day job for each project manager. Integrating risk tasks into that daily routine is the easiest solution. Risk tasks may be carried out to identify or analyze risks or to generate, select and implement responses. The daily effort of integrating risk tasks keeps your project focused on the current situation of risks and helps you stay on top of their relative importance.

Summary

The benefit of risk management in projects is huge because the outcome of project failure is wasted dollars that steal investor profits and have a negative impact on the organization's bottom-line. Risk assessments allow you to deal with uncertain project events in a proactive manner. This allows you to deliver your project on time, on budget and with quality results.

CHAPTER 26

Quality Management

Quality Management ensures that an organization, product or service is consistent. It has **four** main components: quality planning, quality assurance, quality control and quality improvement. Quality management is focused not only on product and service quality, but also on the means to achieve it. Quality management, therefore, uses quality assurance and control of processes as well as products to achieve more consistent quality.

Quality Assurance (QA) -- the maintenance of a desired level of quality in a service or product, especially by means of attention to every stage of the process of delivery or production.

Quality Control (QC) -- a system of maintaining standards in manufactured products by testing a sample of the output against the specification.

What is 'Quality'? The term 'quality has several definitions in regards to business and the professional world. For some, quality means meeting or exceeding customer expectations. Others consider quality to be the standard of something as measured against a defined criteria. And lastly, quality is sometimes referred to as a degree of excellence. None of these definitions are wrong. Each is an accurate statement, and they are only dependent upon the circumstances in which they are used.

Total quality management (TQM) consists of organization-wide efforts to install and make permanent a climate in which an organization continuously improves its ability to deliver high-quality products and services to customers. While there is no widely agreed-upon approach, TQM efforts typically draw heavily on the previously developed tools and techniques of quality control. TQM seeks to improve the quality of products and services through ongoing refinements in response to continuous feedback.

Quality Assurance vs. Quality Control

Quality Assurance is *process* oriented and focuses on defect *prevention*, while **Quality Control** is *product* oriented and focuses on defect *identification.*

To be successful implementing TQM, an organization must concentrate on the eight key elements:

- Ethics
- Integrity
- Trust
- Training
- Teamwork
- Leadership
- Recognition
- Communication

PDCA (**plan–do–check–act**) is an iterative four-step management method used in business for the control and continual improvement of processes and products. It is also known as the Deming circle/cycle/wheel, Shewhart cycle, or control circle/cycle. Just as a circle has no end, the PDCA cycle should be repeated again and again for continuous improvement.

When to Use Plan–Do–Check–Act

- As a model for continuous improvement.
- When starting a new improvement project.
- When developing a new or improved design of a process, product or service.
- When defining a repetitive work process.
- When planning data collection and analysis in order to verify and prioritize problems or root causes.
- When implementing any change.

Plan–Do–Check–Act Procedure

1. **Plan.** Recognize an opportunity or define the problem. Collect relevant data. Ascertain the problem or identify the opportunity. Plan the change or improvement.

2. **Do.** Test the change improvement. Carry out a small-scale study. Decide upon a measurement to gauge its effectiveness. Develop and implement a solution.

3. **Check.** Review the test. Analyze the results. Confirm the results through before-and-after data comparison. Identify what you've learned.

4. **Act.** Document the results. Inform others about process changes or improvement. Take action based on what you learned in the study step: If the change did not work, go through the cycle again with a different plan. If you were successful, incorporate what you learned from the test into wider changes. Use what you learned to plan new improvements, beginning the cycle again. Make recommendations for the problem to be addressed in the next PDCA cycle.

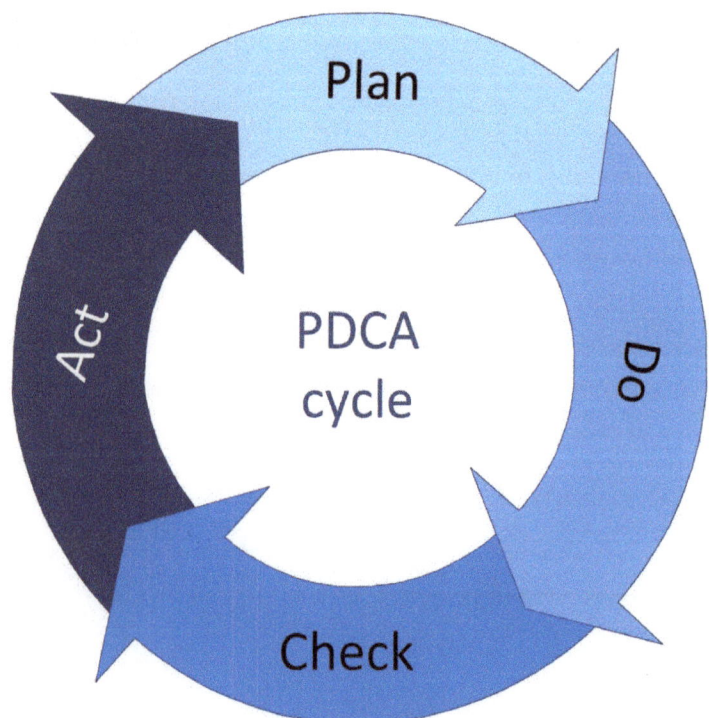

Quality Assurance vs. Quality Control Comparison Chart

	Quality Assurance	Quality Control
Definition	QA is a set of activities for ensuring quality in the processes by which products are developed.	QC is a set of activities for ensuring quality in products. The activities focus on identifying defects in the actual products produced.
Focus on	QA aims to prevent defects with a focus on the process used to make the product. It is a proactive quality process.	QC aims to identify (and correct) defects in the finished product. Quality control, therefore, is a reactive process.
Goal	The goal of QA is to improve development and test processes so that defects do not arise when the product is being developed.	The goal of QC is to identify defects after a product is developed and before it's released.
How	Establish a good quality management system and the assessment of its adequacy. Periodic conformance audits of the operations of the system.	Finding & eliminating sources of quality problems through tools & equipment so that customer's requirements are continually met.
What	Prevention of quality problems through planned and systematic activities including documentation.	The activities or techniques used to achieve and maintain the product quality, process and service.
Responsibility	Everyone on the team involved in developing the product is responsible for quality assurance.	Quality control is usually the responsibility of a specific team that tests the product for defects.
Example	Verification is an example of QA	Validation/Software Testing is an example of QC
Statistical Techniques	Statistical Tools & Techniques can be applied in both QA & QC. When they are applied to processes (process inputs & operational parameters), they are called Statistical Process Control (SPC); & it becomes the part of QA.	When statistical tools & techniques are applied to finished products (process outputs), they are called as Statistical Quality Control (SQC) & comes under QC.
As a tool	QA is a managerial tool	QC is a corrective tool
Orientation	QA is process oriented	QC is product oriented

Six Sigma and Quality Management

'Six Sigma' is a business management strategy which aims at improving the quality of processes by minimizing and eventually removing the errors and variations. The concept of Six Sigma was introduced by Motorola in 1986, but was popularized by Jack Welch who incorporated the strategy in his business processes at General Electric. The concept of Six Sigma came into existence when one of Motorola's senior executives complained of Motorola's bad quality. Bill Smith eventually formulated the methodology in 1986.

Quality plays an important role in the success and failure of an organization. Neglecting an important aspect like quality, will not let you survive in the long run. **Six Sigma ensures superior quality of products by removing the defects in the processes and systems**. Six sigma is a process which helps in improving the overall processes and systems by identifying and eventually removing the hurdles which might stop the organization to reach the levels of perfection. According to sigma, any sort of challenge which comes across in an organization's processes is considered to be a defect and needs to be eliminated.

Organizations practicing Six Sigma create special levels for employees within the organization. Such levels are called as: "Green belts", "Black belts" and so on. Individuals certified with any of these belts are often experts in six sigma process. **According to Six Sigma any process which does not lead to customer satisfaction is referred to as a defect and has to be eliminated from the system to ensure superior quality of products and services**. Every organization strives hard to maintain excellent quality of its brand and the process of six sigma ensures the same by removing various defects and errors which come in the way of customer satisfaction.

The process of Six Sigma originated in manufacturing processes but now it finds its use in other businesses as well. Proper budgets and resources need to be allocated for the implementation of Six Sigma in organizations.

Following are the two Six Sigma methods:
- DMAIC
- DMADV

DMAIC focuses on improving existing business practices. DMADV, on the other hand focuses on creating new strategies and policies.

DMAIC Method:

D - **Define the Problem**. In the first phase, various problems which need to be addressed to are clearly defined. Feedbacks are taken from customers as to what they feel about a particular product or service. Feedbacks are carefully monitored to understand problem areas and their root causes.

M - **Measure and find out the key points of the current process**. Once the problem is identified, employees collect relevant data which would give an insight into current processes.

A - **Analyze the data**. The information collected in the second stage is thoroughly verified. The root cause of the defects are carefully studied and investigated as to find out how they are affecting the entire process.

I - **Improve the current processes** based on the research and analysis done in the previous stage. Efforts are made to create new projects which would ensure superior quality.

C - **Control the processes** so that they do not lead to defects.

DMADV Method:

D - Design strategies and processes which ensure hundred percent customer satisfaction.

M - Measure and identify parameters that are important for quality.

A - Analyze and develop high level alternatives to ensure superior quality.

D - Design details and processes.

V - Verify various processes and finally implement the same.

Six Sigma is a quality program that, when all is said and done, improves your customer's experience, lowers your costs, and builds better leaders.

Importance of Quality Management

Quality management ensures superior quality products and services. Quality of a product can be measured in terms of performance, reliability and durability. Quality is a crucial parameter which differentiates an organization from its competitors. Quality management tools ensure changes in the systems and processes which eventually result in superior quality products and services. Quality management methods such as Total Quality management or Six Sigma have a common goal - to deliver a high quality product. Quality management is essential to create superior quality products which not only meet but also exceed customer satisfaction. Customers need to be satisfied with your brand. Business marketers are successful only when they emphasize on quality rather than quantity. Quality products ensure that you survive the cut throat competition with a smile.

Quality management is essential for customer satisfaction which eventually leads to customer loyalty. How do you think businesses run? Do businesses thrive only on new customers? It is important for every business to have some loyal customers. You need to have some customers who would come back to your organization no matter what.

Are you likely to purchase a Visio television again if you experienced a lot of problems with your current Visio television? For most people, the answer is NO.

Customers would return to your organization only if they are satisfied with your products and services. Make sure the end-user is happy with your product. Remember, a customer would be happy and satisfied only when your product meets his expectations and fulfills his needs. Understand what the customer expects from you? Find out what actually his need is? Collect relevant data which would give you more insight into customer's needs and demands. Customer feedbacks should be collected on a regular basis and carefully monitored. Quality management ensures high quality products and services by eliminating defects and incorporating continuous changes and improvements in the system. High quality products in turn lead to loyal and satisfied customers who bring ten new customers along with them. Do not forget that you might save some money by ignoring quality management processes but ultimately lose out on your major customers, thus incurring huge losses. Quality management ensures that you deliver products as per promises made to the customers through various modes of promotions. **Quality management tools help an organization to design and create a product which the customer actually wants and desires**.

Quality Management ensures increased revenues and higher productivity for the organization. Remember, if an organization is earning, employees are also earning. Employees are frustrated only when their salaries or other payments are not released on time. Yes, money is a strong motivating factor. Would you feel like working if your organization does not give you salary on time? Ask yourself. Salaries are released on time only when there is free cash flow. Implementing quality management tools ensure high customer loyalty, thus better business, increased cash flow, more content employees, and a more positive workplace. Quality management processes make the organization a better place to work.

Remove unnecessary processes which merely waste employee's time and do not contribute much to the organization's productivity. Quality management enables employees to deliver more work in less time.

Quality management helps organizations to reduce waste and inventory. It enables employees to work closely with suppliers and incorporate a more efficient operating philosophy.

Quality management ensures close coordination between employees of an organization. It inculcates a strong feeling of team work in the employees.

Quality Management Tools

Quality Management tools help organization collect and analyze data for employees to easily understand and interpret information. Quality Management models require extensive planning and collecting relevant information about end-users. Customer feedbacks and expectations need to be carefully monitored and evaluated to deliver superior quality products.

Quality Management tools help employees identify the common problems which are occurring repeatedly and also their root causes. Quality Management tools play a crucial role in improving the quality of products and

services. With the help of Quality Management tools employees can easily collect the data as well as organize the collected data which would further help in analyzing the same and eventually come to concrete solutions for better quality products.

Quality Management tools make the data easy to understand and enable employees to identify processes to rectify defects and find solutions to specific problems.

Following are the quality management tools:

* **Check List** - Check lists are useful in collecting data and information easily .Check list also helps employees to identify problems which prevent an organization to deliver quality products which would meet and exceed customer expectations. Check lists are nothing but a long list of identified problems which need to be addressed. Once you find a solution to a particular problem, tick it immediately. Employees refer to check list to understand whether the changes incorporated in the system have brought permanent improvement in the organization or not?

* **Pareto Chart** - The credit for Pareto Chart goes to Italian Economist - Wilfredo Pareto. A Pareto chart is a type of chart that contains both bars and a line graph, where individual values are represented in descending order by bars, and the cumulative total is represented by the

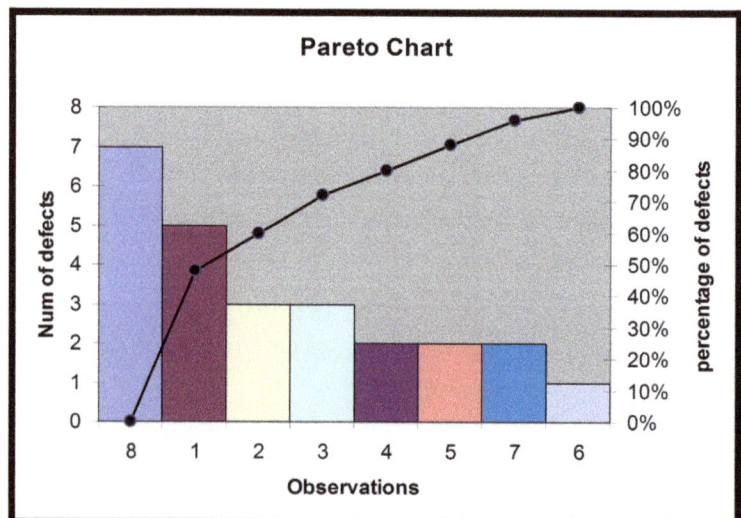

FIGURE 5. Pareto Chart

line. Pareto Chart helps employees to identify the problems, prioritize them and also determine their frequency in the system. Pareto Chart often represented by both bars and a line graph identifies the most common causes of problems and the most frequently occurring defects. Pareto Chart records the reasons which lead to maximum customer complaints and eventually enables employees to formulate relevant strategies to rectify the most common defects.

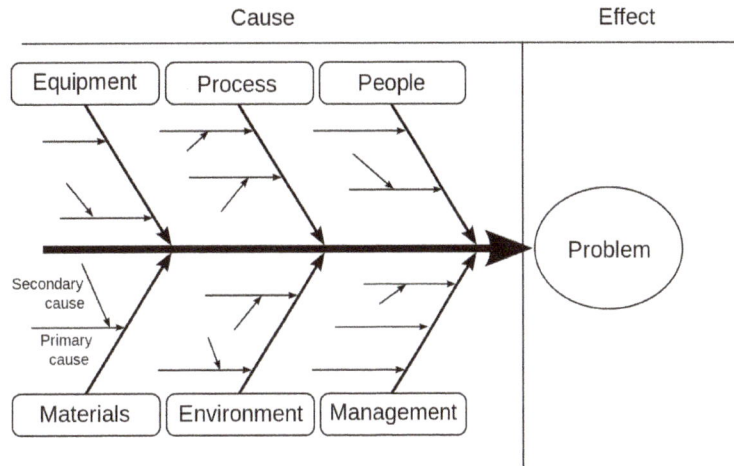

FIGURE 6. *Cause and Effect Diagram*

- **The Cause and Effect Diagram** - Also referred to as "Fishbone Chart" (because of its shape which resembles the side view of a fish skeleton) and Ishikawa diagrams after its creator Kaoru Ishikawa, Cause and Effect Diagram records causes of a particular and specific problem .The cause and effect diagram plays a crucial role in identifying the root cause of a particular problem and also potential factors which give rise to a common problem at the workplace.

- **Histogram** - Histogram, introduced by Karl Pearson is nothing but a graphical representation showing intensity of a particular problem or variable. Histogram helps identify the cause of problems in the system by the shape as well as width of the distribution.

- **Scatter Diagram** - Scatter Diagram is a quality management tool which helps to analyze relationship between two variables. In a scatter chart, data is represented as points, where each point denotes a value on the horizontal axis and vertical axis.

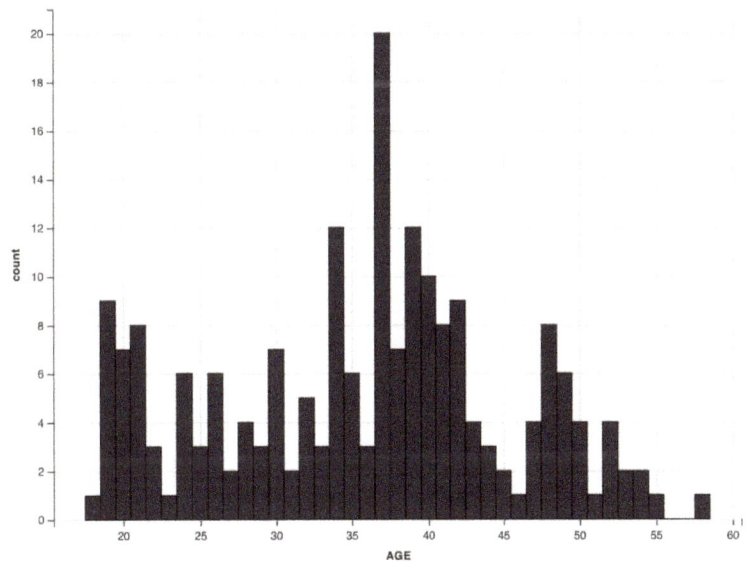

FIGURE 7. *Histogram*

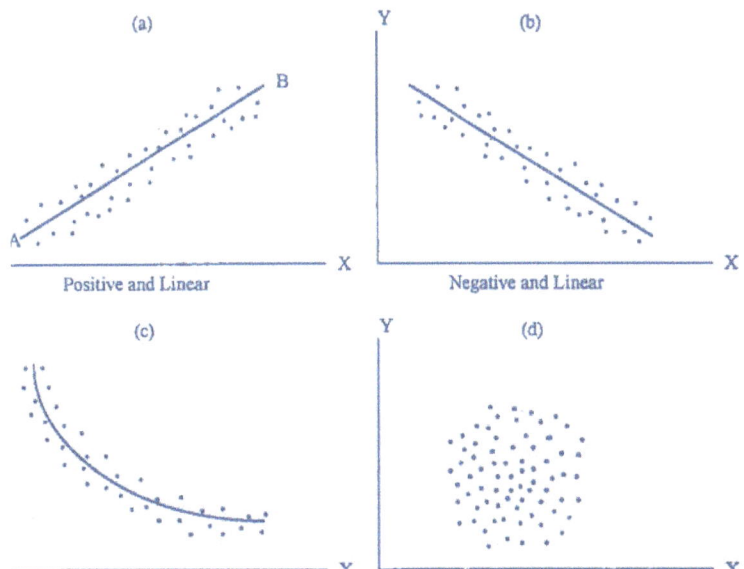

FIGURE 8. *Scatter Diagram*

255

Scatter Diagram shows many points which show a relation between two variables.

- **Graphs** - Graphs are the simplest and most commonly used quality management tools. Graphs help to identify whether processes and systems are as per the expected level or not and if not also record the level of deviation from the standard specifications.

Benefits of a Quality Management

Like any business decision, there has to be some good reasons for implementing a Quality Management System (QMS). Some people think that total quality management and quality management systems are a never-ending expense and a time-drain away from other revenue producing activities. However, a proper QMS delivers real value and benefits to the organization; and not only makes it worthwhile, but a fruitful endeavor. Following are a some of the major benefits to having an effective QMS.

A Quality Management System Improves Organizational Effectiveness

Successful organizations have certain goals they are pursuing in every segment of the organization, including customer service, sales, fulfillment of orders, and purchasing to name a few. How well is your business doing in meeting goals in all segments of your organization? Frequently, organizations will claim to have established priorities to fulfill; yet without actively measuring performance so they do not have a clue as to how effective they are in these critical areas.

An essential role of a QMS is establishing important goals for critical operational sectors, then measuring and reviewing the organization's effectiveness at reaching those goals. When an organization monitors the performance relative to these goals (as well as action plans to close gaps), they become more effective.

A Quality Management System Improves Customer Satisfaction

Most businesses will not survive for long if they have unfavorable customer satisfaction results. How well is your business doing at delivering what customers want and expect?

A properly designed and implemented QMS makes customer satisfaction a high priority. It requires that customer feedback be encouraged from multiple sources, and that the information collected be managed and used to improve customer satisfaction.

A Quality Management System Improves Compliance

Many organizations report that compliance to standards and regulations is one of their biggest challenges. Many have difficulty conforming to the requirements.

Compliance to standards and regulations can be built into the QMS, including into the documentation system, the internal auditing program, and the measurement and reporting system. Compliance is much less of a challenge when it is integrated into a functioning QMS.

A Quality Management System Improves Organizational Culture and Better Employee Moral

Employees at many organizations report that one of the most frustrating aspects of their job is that they are unsure of the organizational objectives and priorities; and their role in fulfilling them is somewhat ambiguous. This is particularly true when the priorities are constantly shifting. This creates a negative culture of uncertainty, fear, and resentment.

A QMS requires that an organization's goals and priorities are clearly established, communicated, and aligned with operational activities. This gives employees confidence that they are doing the right thing, and that the right thing remains consistent. This increases productivity by promoting a much more positive work atmosphere and higher morale.

A Quality Management System Improves Documentation

Poor documentation management is a common problem for many organizations. This puts organization at risk and limits success. A QMS helps to define document management systems and document needs so activities and processes are clearly recorded and available.

A Quality Management System Increases Revenue

Companies which maintain an effective QMS experience improved financial performance and increased productivity when compared to companies where QMS is a low priority.

A Quality Management System Increases Efficiency

QMS forces an organization to think about their operations and actions. Creating guidelines for tasks that are undertaken daily, or regularly, improves efficiency. Establishing training templates and plans further improves efficiency. There is solid proof that having a QMS in place helps a business to run more effectively.

Quality Management Systems Leads to Better Decisions

Key business decisions shouldn't be based on intuition or a "gut feeling". Having proper procedures and structure in place ensure that decisions are made based on rationality and logic; as well as an understanding of what has happened in the past. Having the ability to make better decisions enables an organization to make continuous improvements and improve on what they offer.

A Quality Management System Improves Relations between Suppliers and Vendors

Flourishing companies understand that their relationship with other parties in the supply chain impacts their overall level of success. QMS aids in integrating a level of mutually beneficially success into relationships. A good QMS provides a platform to evaluate suppliers and then to carry out audits of the ongoing work being undertaken, or products being acquired, by the selected vendors.

A Quality Management System Improves Consistency

One of the biggest factors contributing to the success of an organization derived from an effective QMS is that it helps to bring consistency to the workplace. A business that is able to operate in the same manner on a consistent basis will be more efficient, improve their output and give customers, clients and interested parties a greater level of confidence in what they provide and offer. Companies that operate consistently are more likely to achieve and maintain success.

Conclusion

A properly executed quality management system positively affects every aspect of an organization's performance. There are a wide variety of helpful tools available to help an organization effectively operate an quality management system efficiently. The advantages of a quality management system for your business, as described above, are numerous. Quality improvement not only benefits the organization, customers, and business relationships, but also offers a variety of problem-solving techniques and strategic components to better equip the organization to successfully transition into the future. Successful organizations and effective leaders realize that quality management isn't an ongoing cumbersome aggravation or a burdensome expense, but a valuable investment that delivers positive results and substantial rewards.

PART IV

Appendices

Resources

Managing Your Business

- The IRS has remarkably clear publications on farming tax issues. They can be found at www.irs.gov, or request the "Farmer's Tax Guide" from 800-829-3676.
- For information about farmland property tax assessments, the American Farmland Trust and the USDA have compiled numerous publications on the website www.farmlandinfo.org.
- How to track electricity usage: The Kill A Watt power measurement tool is available from Real Goods. 888-567-6527; www.realgoods.com.
- Deciding on a legal structure for your business requires the advice of an attorney or accountant. Here is a link to a publication from Kansas State University that describes the various options: www.ksre.ksu.edu/bookstore/pubs/MF2696.pdf.
- **C12 Group**: It's an objective advisory board for brainstorming and decision making, learning what you don't know and focusing on areas you need to sharpen. In a confidential, non-competing trusted C12 peer board, you'll learn from your group's wisdom and insight — and encourage and hold each other accountable to the principles and core values that guide you. http://www.c12group.com

Insurance

- The company that provides our farmers cooperative's products liability coverage (as well as our personal farm policy) is called Goodville Mutual Casualty Company, www.goodville.com. It's based in New Holland, Pennsylvania, and covers a lot of direct-market farmers in these nine states: Pennsylvania, Delaware, Maryland, Virginia, Ohio, Indiana, Illinois, Kansas, and Oklahoma. You can call the company at 717-354-4921 to find an agent near you who sells their policies.
- InterWest Insurance Services, in Sacramento, California. 800444-4134; www.iwins.com.

Payroll services

- QuickBooks Payroll, www.intuit.com.
- ADP. 800-225-5237; www.adp.com.
- Local accounting offices also do payroll for small businesses; be sure to compare prices and services.
- Your state's labor department.

FARM YOUR SPACE

The author is creating a website **www.FarmYourSpace. com** that will have a great deal of helpful information on it, a Q&A feature where you can get your questions answered, many articles, pics and videos related to aquaponics, vertical gardening, and other helpful ideas to maximize your space; as well as improving the effectiveness and efficiency of your farming operation. I encourage you to check it out. Table 1 lists some of the topics FarmYourSpace.com will cover:

TABLE 1.

	"FARMYOURSPACE" WEBSITE CATEGORIES
1	Aquaponics
2	Animals, Poetry, Livestock
3	Community Forum & Expert Advice
4	Do-It-Yourself
5	Fruit & Nut Trees
6	Healthy Soil, Natural Fertilizers, Natural Pesticides
7	Hydroponics
8	Maximizing Your Space
9	Money Saving Tips & Money Making Ideas
10	Nutrition, Health, Organics
11	Raised Beds
12	Surviving Regulations & Big Brother (We will keep you informed and petition for your rights on related government policy and legislation issues.)
13	Sustainable Farming
14	Traditional Gardening & New Ideas!!!
15	Vertical Gardening
16	Water
17	Living off the grid.
18	Cooking
19	Survival
20	WHAT IS WRONG WITH THE WORLD (issues related to Pesticides / Herbicides / Fungicides, GMOs, EMF, Radiation, Pollution, Water contamination, Destruction of natural habitat, Food Contaminants, Drinking Water Contaminants / Fluoride, Health Problems Increasing, Political Corruption, Dumbing Down of the Public, Etc.

RELIABLE ALTERNATIVE NEWS SOURCES

Approximately 90 percent of all news media outlets in the USA are owned by a total of only six companies. The owners and/or head of all these companies have been reported to have ties to the Bilderberg Group. Bilderberg is a highly secretive, elitist, international think tank and policy-forming group. This globalist establishment of government leaders and media company heads work together to execute a covert globalist agenda that only benefits them and is in opposition to the best interest of the common person.

In addition advertisers support media companies with millions of dollars in ads, and that support comes with strings attached. Companies spending such large sums of money on advertisements have no qualms about requiring media outlets to implement policy measures that will sway public opinion as a stipulation for receiving those advertising dollars.

The Obama administration dished out tremendous amounts of funding, grants, and special favors to mainstream media companies and NPR. These government funds were dispersed with the understanding that the receiving media outlets would help promote or discourage various issues.

The CIA's influence, and often outright control, of the media began in the 1950s with Operation Mockingbird and has carried through to the present. They have a close alliance with the Council on Foreign Relations which includes mainstream journalists, many of whom are reportedly former CIA agents. As Richard Harwood in his 1993 article "Ruling Class Journalist" stated these journalists, "do not merely analyze and interpret foreign policy; they help make it."

Lastly, foreign companies have purchased and/or substantially invested in media companies and Hollywood studios. These are multi-million dollar transactions. China, widely known for strict censorship and exclusive propaganda media messaging, is one of the largest foreign owners of American media entertainment companies.

As a result of all the above forces, it is next to impossible to obtain unbiased reporting from the mainstream media. Mainstream media is pressured to the point where the truth gets buried under a scripted narrative resulting in propaganda, fake news, and twisted truth. Bottom line: Mainstream news is corrupt beyond measure.

Public manipulation through mainstream media is accomplished by burying stories, over emphasizing stories, reporting with bias, practicing subtle deceit and flat out lying. This public persuasion is used to push an agenda.

In China, a person's social credit score goes down when citizens question authority or communicate anything that does not abide with the establishment. A low social credit score in China has serious implications. Since its recent inception, millions of Chinese people have already been barred from air transport and stopped from buying high-speed train tickets. Additional consequences are to be implemented in the near future.

As of this writing in 2018, similar types of measures are beginning to take shape in Europe and on social media. Google, Facebook, YouTube, and Twitter are reported to have similar mechanisms for rating and penalizing users with viewpoints which oppose their agenda. Some of the penalties include, demonetizing sites, altering 'likes', reducing viewership, and in some

instances even direct censorship and terminating accessibility.

Free speech is being labeled as offensive or hate speech, if it doesn't fit the establishment's narrative. Anyone with an opposing view is often attacked personally, rather than being engaged in a debate over the counter view point. Whether liberal or conservative, this suppression of free speech and ideas should greatly concern everyone.

Tyranny operates best when the minds of the public are won over. Towards this end, mainstream media is attempting to mold the population via a wealth of controlled information.

To obtain more accurate reporting of the news one must find and depend upon other media sources which don't have any or at least as many strings attached. The remaining 10 percent of media outlets that are independent -- although not perfect -- tend to provide more reliable information and can be better trusted. Several of these valuable independent media outlets are listed on the next page.

NOTE: Although the author recommends these resources as a means of obtaining more unbiased information, the author is not in agreement with any profanity, fowl language, emotional rants, or condemnation of a person that may be presented on these or any other media outlets. Condemnation of wrong behavior is acceptable, but not condemnation of a person.

- www.Drudge.com
- www.NatrualNews.com
- www.DailyCaller.com
- www.InfoWars.com
- www.organicconsumers.org
- www.fluoridealert.org
- www.NaturalSociety.com
- www.GeoEngineeringWatch.org
- www.fluoridealert.org/articles/50-reasons (50 reasons why fluoride in H2O is bad)

The following channels are also known to provide more accurate reporting and unbiased narratives:
- www.REAL.video (a new platform alternative to YouTube censorship)
- Paul Joseph Watson, www.youtube.com/user/PrisonPlanetLive
- Mike Dice, www.youtube.com/user/MarkDice
- Black Pigeon Speaks, https://www.youtube.com/user/TokyoAtomic/videos
- Lauren Southern, https://www.youtube.com/channel/UCla6APLHX6W3FeNLc8PYuvg
- Stefan Molyneux, www.youtube.com/user/stefbot

APPENDIX 2

Several of the author's previous books.

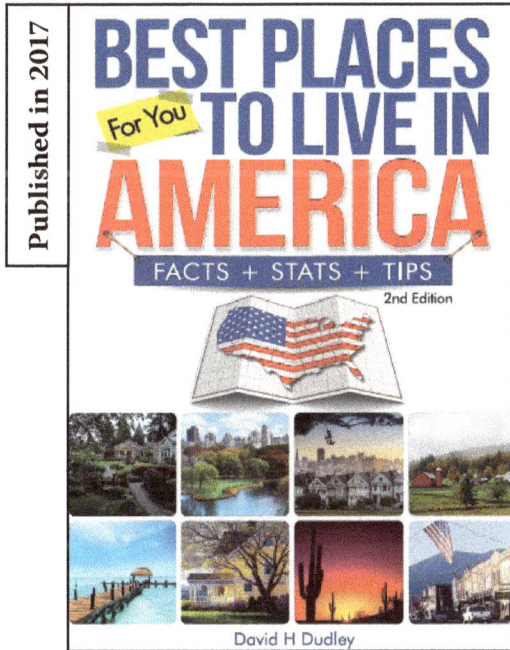

Published in 2017

BEST PLACES
For You
TO LIVE IN
AMERICA
FACTS + STATS + TIPS
2nd Edition

David H Dudley

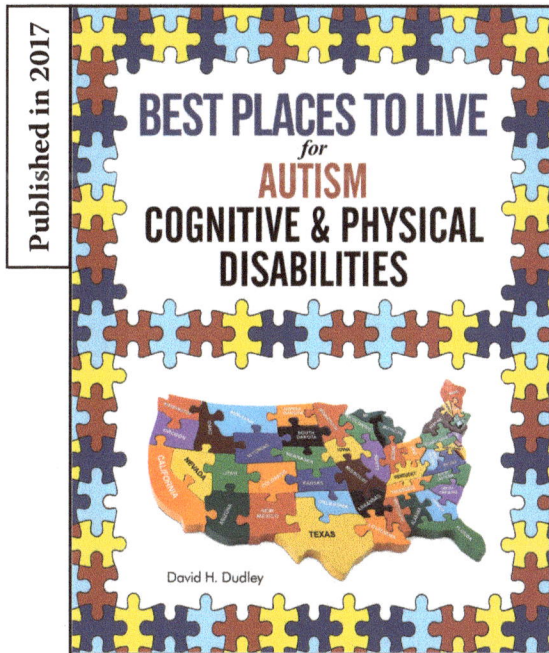

Published in 2017

BEST PLACES TO LIVE
for
AUTISM
COGNITIVE & PHYSICAL
DISABILITIES

David H. Dudley

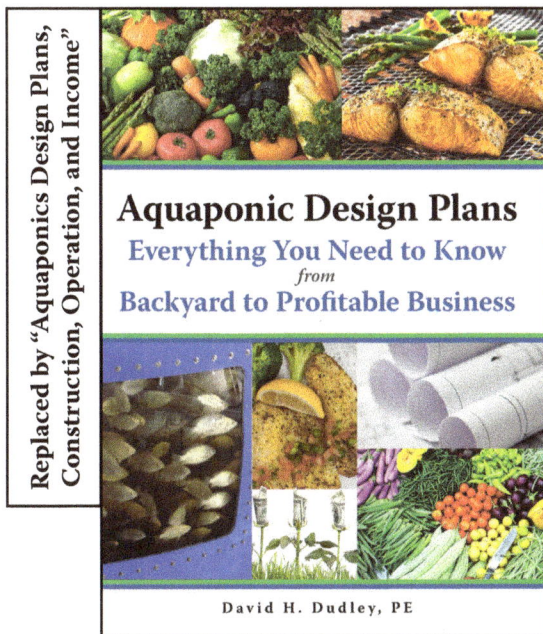

Replaced by "Aquaponics Design Plans, Construction, Operation, and Income"

Aquaponic Design Plans
Everything You Need to Know
from
Backyard to Profitable Business

David H. Dudley, PE

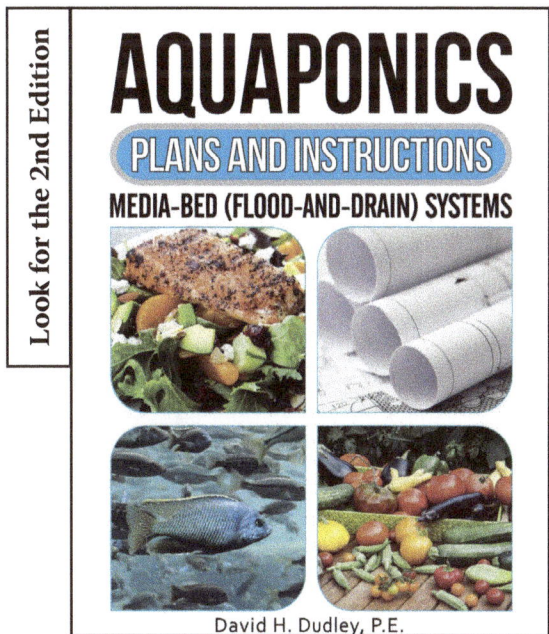

Look for the 2nd Edition

AQUAPONICS
PLANS AND INSTRUCTIONS
MEDIA-BED (FLOOD-AND-DRAIN) SYSTEMS

David H. Dudley, P.E.

AQUAPONICS
HOW TO DO EVERYTHING
2nd Edition

FROM BACKYARD TO
PROFITABLE BUSINESS

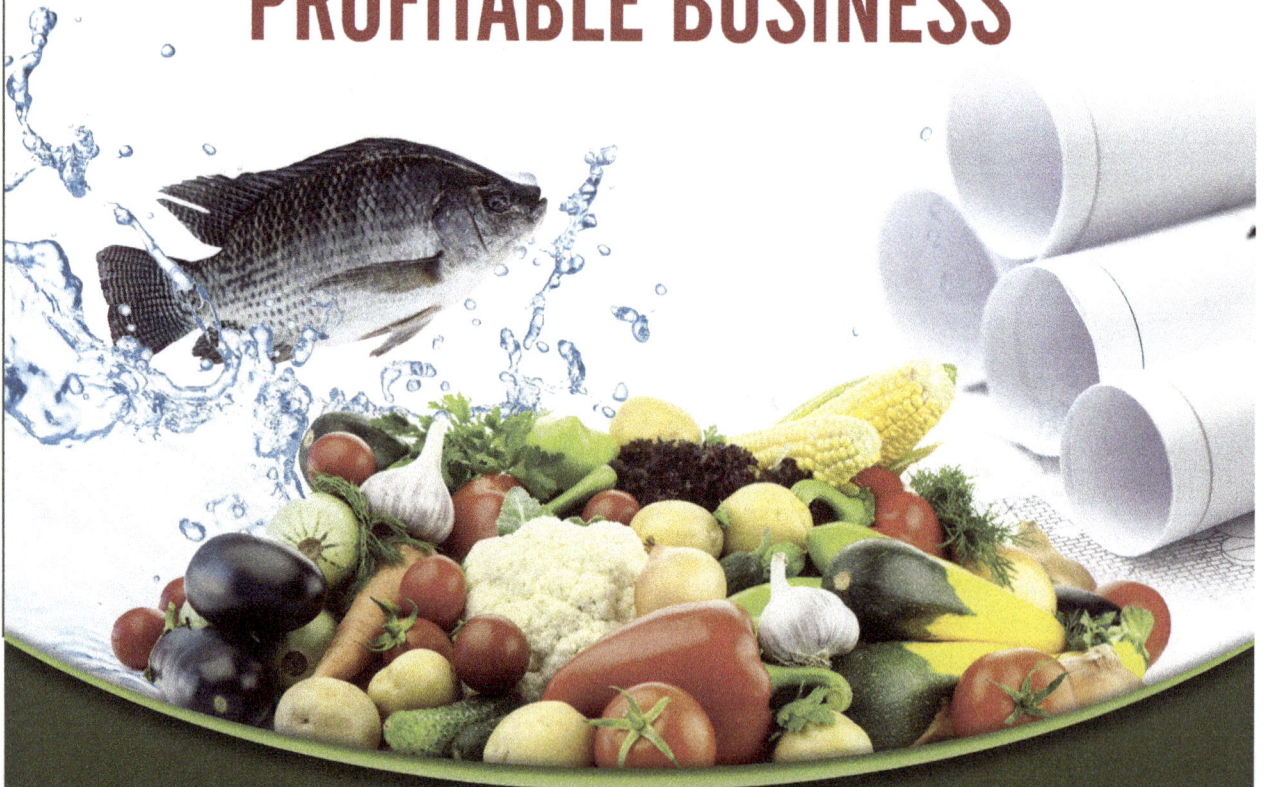

David H. Dudley, PE

AQUAPONICS FOR PROFIT

EARN **EXTRA MONEY** OR CREATE A **SUCCESSFUL COMMERCIAL BUSINESS**

David H. Dudley, PE

Aquaponics for Profit

- What are the best methods for making money with aquaponics?
- Which edible aquaponic fish species will generate the most revenue per pound?
- Which aquaponic vegetables provide the highest profit margin?
- Which fish and vegetables are in highest demand by consumers?
- Do you know about all of the non-vegetable plants that can provide you with greater revenue than vegetable plants?
- What costs are involved in setting-up and operating an aquaponic system?
- What is the cost-benefit analysis of an aquaponic system?
- Where and how can I sell my aquaponic harvest?
- What regulations and legalities are involved in setting-up and operating an aquaponic business?
- Did you know that there are other aquatic species that can be grown in aquaponics which can generate more revenue than the commonly grown edible aquaponic fish species?
- What is the best way to barter my aquaponic harvest?
- How can I get my products officially labeled as 'organic'?
- What is the best approach to having a successful aquaponics business that will produce the largest profit margin?
- How can I earn extra money with my small backyard aquaponic system?
- Which type of aquaponic system – Media-Bed/Flood-and-Drain, Nutrient Film Technique (NFT), Raft System / Deep Water Culture (DWC) – is the most profitable?
- What are the pros and cons to each of these types of aquaponic systems?
- In addition to selling your harvest, are you aware of all the other ways in which you can earn revenue from your aquaponic system?
- How much time would I need to invest to have an aquaponic system that will feed my family and provide us with some extra income?

This user-friendly easy read book will answer the above a questions. This valuable resource is also packed with the necessary information that will not only show you how to make extra money with your aquaponic system, but to grow it into a successful commercial business; if that is your desire. **Also included are two real-world aquaponic business plans.** This book is an excellent investment that will reward you greatly with the knowledge needed to earn extra money through aquaponics or optimize revenue from a commercial aquaponic operation.

AQUAPONICS FOR PROFIT

EARN **EXTRA MONEY** OR CREATE A **SUCCESSFUL COMMERCIAL BUSINESS**

What are the best methods for making money with aquaponics? Which edible aquaponic fish species will generate the most revenue per pound? Which aquaponic vegetables provide the highest profit margin? Which fish and vegetables are in highest demand by consumers? Do you know about all of the non-vegetable plants that can provide you with greater revenue than vegetable plants? What costs are involved in setting-up and operating an aquaponic system? What is the cost-benefit analysis of an aquaponic system? Where and how can I sell my aquaponic harvest? What regulations and legalities are involved in setting-up and operating an aquaponic business? Did you know that there are other aquatic species that can be grown in aquaponics which can generate more revenue than the commonly grown edible aquaponic fish species? What is the best way to barter my aquaponic harvest? How can I get my products officially labeled as 'organic'? What is the best approach to having a successful aquaponics business that will produce the largest profit margin? How can I earn extra money with my small backyard aquaponic system? Which type of aquaponic system – Media-Bed/Flood-and-Drain, Nutrient Film Technique (NFT), Raft System / Deep Water Culture (DWC) – is the most profitable? What are the pros and cons to each of these types of aquaponic systems? In addition to selling your harvest, are you aware of all the other ways in which you can earn revenue from your aquaponic system? How much time would I need to invest to have an aquaponic system that will feed my family and provide us with some extra income?

This user-friendly easy read book will answer the above a questions. This valuable resources is also packed with the necessary information that will not only show you how to make extra money with your aquaponic system, but to grow it into a successful commercial business; if that is your desire. Also included are two real-world aquaponic business plans. This book is an excellent investment that will reward you greatly with the knowledge needed to earn extra money through aquaponics or optimize revenue from a commercial aquaponic operation.

David H. Dudley is a professional aquaponics consultant who has helped many individuals and companies develop aquaponics systems. His accomplished career in aquaponics, hydroponics, and aquaculture includes serving as the Construction Manager of the Oklahoma Aquarium, Engineering Manager of the nation's largest caviar producing company, overseeing life support systems of four large aquaculture facilities, designing a $5M aquaculture operation for white sturgeon, and Project Manager of a large fishing clinic facility for the U.S. Department of Wildlife. David also holds advanced degrees in civil engineering and nutrition/dietetics, owns a commercial nursery, and has several decades of experience in vegetable gardening. David understands every facet of aquaponics and clearly communicates aquaponics in a way that truly helps others.

www.FarmYourSpace.com

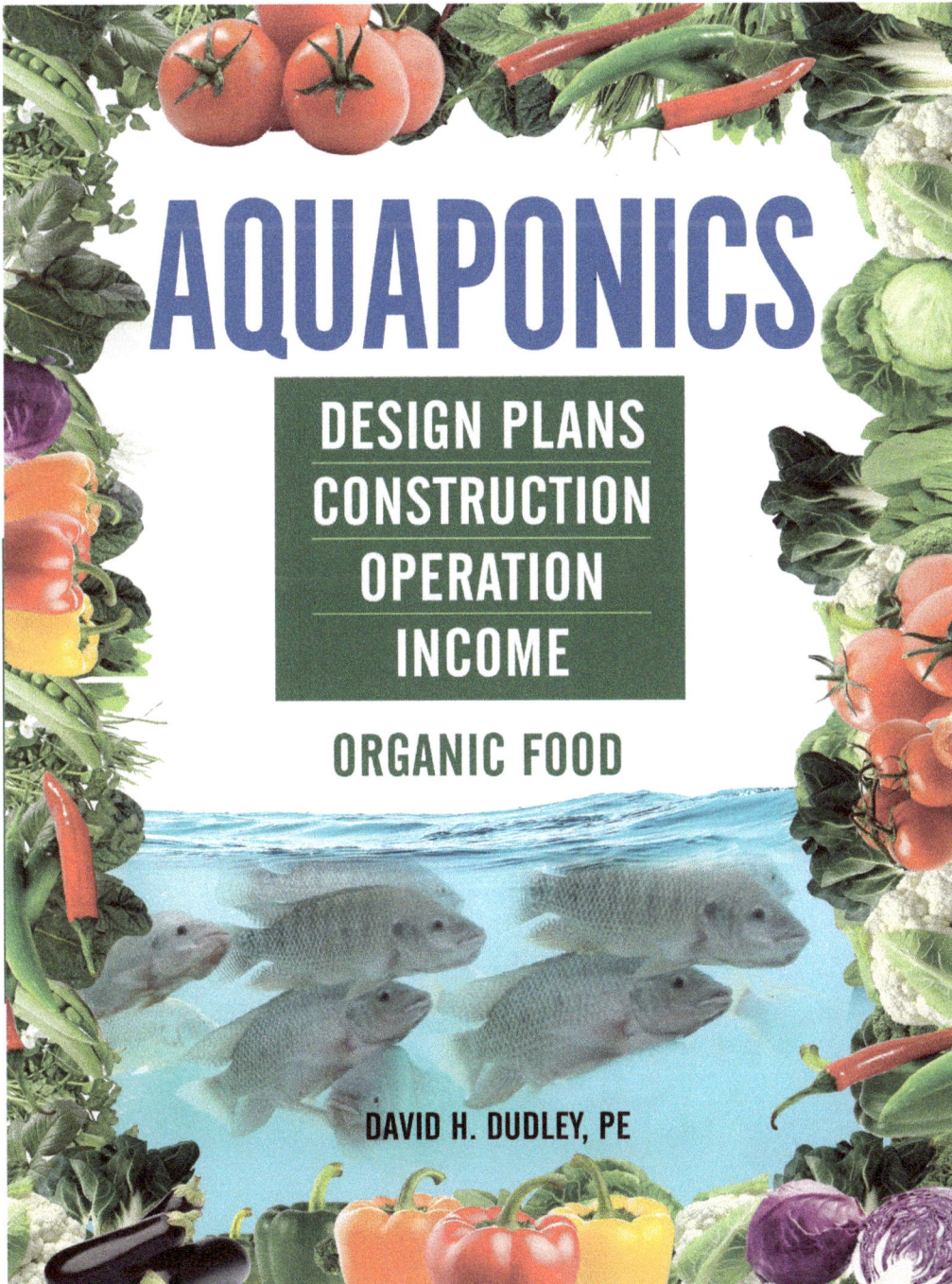

AQUAPONICS

DESIGN PLANS
CONSTRUCTION
OPERATION
INCOME

ORGANIC FOOD

GROW YOUR OWN HEALTHY FOOD SAVE MONEY EARN EXTRA INCOME

BEGINNER TO COMMERCIAL SMALL TO LARGE AQUAPONIC SYSTEMS

This 625+ page easy to follow comprehensive book provides illustrations, step-by-step instructions, and real life photos showing you how to do everything.

This 625+ page user-friendly book (how-to-guide) provides:

✓ EVERYTHING YOU NEED TO KNOW ABOUT AQUAPONICS
✓ HOW TO SET-UP & OPERATE AN AQUAPONIC SYSTEM OF ANY SIZE
✓ HOW TO OBTAIN FINANCIAL REWARDS FROM YOUR AQUAPONICS SYSTEM

Within these sections you will be provided everything you need to know, in an understandable, easy to follow approach; so that you can enjoy environmentally-friendly sustainable farming, consistently feed your family plentiful healthy organic lost cost food, and earn as much extra income as desired. Best of all, this book will show you how to accomplish these objectives in the most efficient way possible. Aquaponics truly is a worthwhile and rewarding endeavor.

The author, **David H. Dudley, P.E.,** is a professional aquaponics consultant who has helped many individuals and companies develop aquaponics systems. His accomplished career in aquaponics, hydroponics, and aquaculture includes serving as the Construction Manager of the Oklahoma Aquarium, Engineering Manager of the nation's largest caviar producing company, overseeing life support systems of four large aquaculture facilities, designing a $5M aquaculture operation for white sturgeon, and Project Manager of a large fishing clinic facility for the U.S. Department of Wildlife.

David also holds advanced degrees in civil engineering and nutrition/dietetics, operated a small commercial nursery for 10+ years, and has several decades of experience in vegetable gardening. David understands every facet of aquaponics and clearly communicates aquaponics in a way that truly helps others.

www.FarmYourSpace.com

272

APPENDIX 5

Country Breeze Farm

A Healthy Living Community for Women with Special Needs

Mission

Country Breeze Farm is a vibrant non-profit community for adult women with cognitive disabilities, and often, other challenges. We are dedicated to enabling our residents to achieve their full potential, providing them and their families a future of hope. Country Breeze Farm encompasses a safe and comfortable rural environment with programs and services that enables our special ladies to be blessed with a good quality life with purpose, love, and nurturing companionship.

Overview

Located in beautiful East Tennessee, our country setting provides residents with a place to be at peace, live a healthy lifestyle, feel at home, build positive relationships and lead rewarding lives. The charming secure campus includes aquaponic greenhouses for growing organic vegetables and flowers, a therapeutic horseback riding arena, chicken houses, fish ponds with a dock, a large gathering center and a multi-purpose building for events, outdoor gardening, a fitness facility, crafting, music programs, indoor/outdoor basketball courts, bocce ball courts, gazebo for relaxing, outdoor shelter for picnics, and nature paths throughout.

Our full-time female staff help our special ladies discover and enjoy occupational projects, whether it be caring for the land and animals, cultivating and harvesting organic produce, helping in the office, or working in other ways on campus. The days start with a community meal and prayer, singing and an encouraging message. This naturally leads to a healthy social atmosphere augmented with lively get-togethers, community outings, excursions into town, and enjoyable work.

Community partnerships with local churches, businesses, schools, colleges, and volunteers play an important role in providing additional opportunities. Our trained staff, security measures in place, and strict operational protocol ensure that our special ladies are always safe and well cared for.

www.CountryBreezeFarm.org

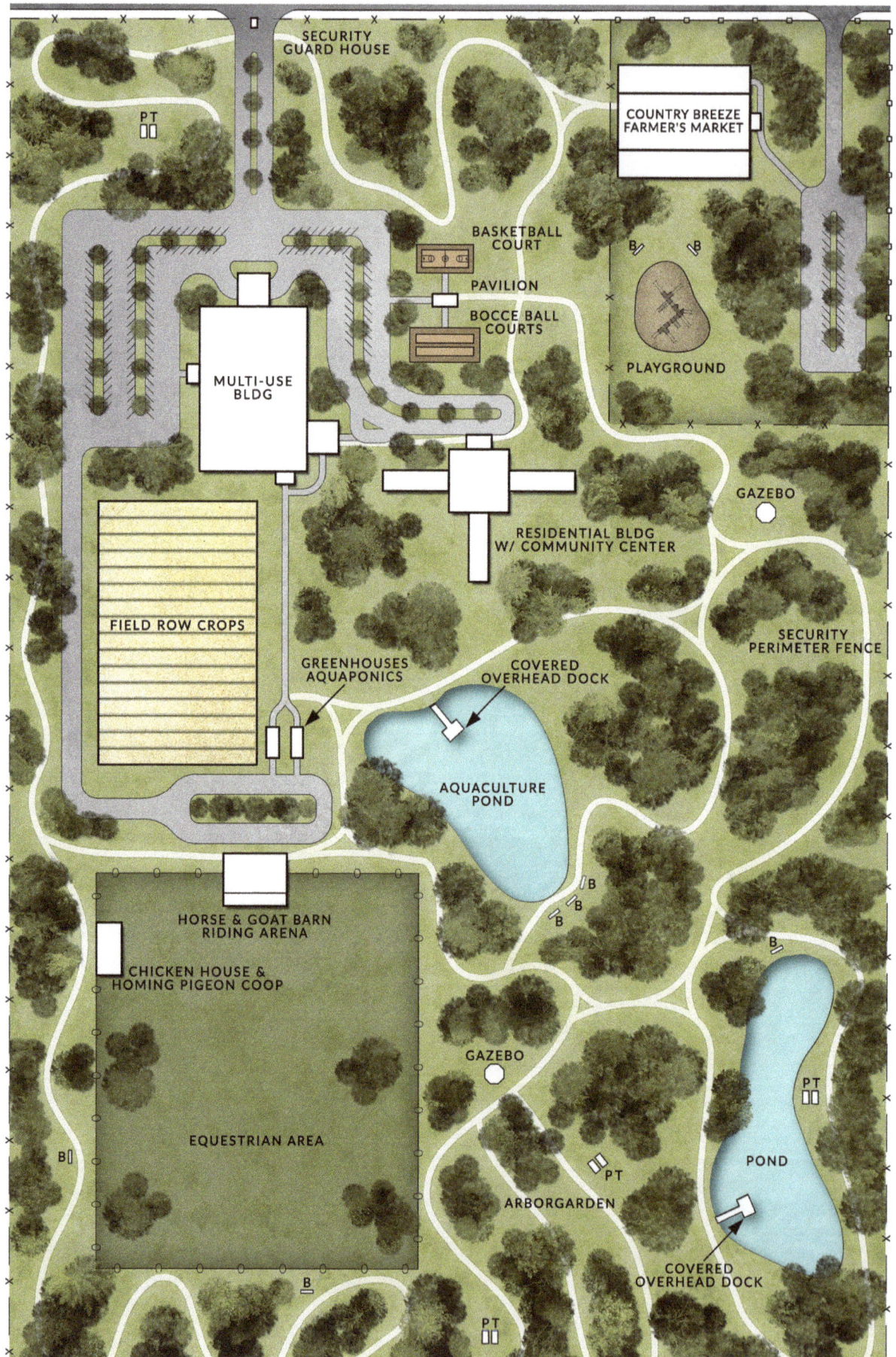

SECURITY
GUARD HOUSE

COUNTRY BREEZE
FARMER'S MARKET

PT

B   B

BASKETBALL
COURT

PAVILION

BOCCE BALL
COURTS

PLAYGROUND

MULTI-USE
BLDG

GAZEBO

RESIDENTIAL BLDG
W/ COMMUNITY CENTER

SECURITY
PERIMETER FENCE

FIELD ROW CROPS

GREENHOUSES
AQUAPONICS

COVERED
OVERHEAD DOCK

AQUACULTURE
POND

B
B
B

B

HORSE & GOAT BARN
RIDING ARENA

CHICKEN HOUSE &
HOMING PIGEON COOP

GAZEBO

PT

PT

EQUESTRIAN AREA

ARBORGARDEN

POND

B

B

COVERED
OVERHEAD DOCK

PT

Country Breeze Farm
A Healthy Living Community for Women with Special Needs

Secure Campus & Facilities

Country Breeze Farm has a wide variety of facilities and infrastructure to support our special ladies. First, consider that the farm has a completely secure campus featuring a security station at the entrance and a secure perimeter fence. The security station has full gate control and provides video monitoring for anyone entering or leaving the premises. In addition to our all female trained staff, video monitoring is provided at various places around the farm, which are accessible to management at all times. Some of the live video feeds can be viewed by families online.

Residential Housing with Community Areas

The living arrangements feature individualized bedrooms custom furnished for each resident according to her special needs. The open community room has tables, couches, and recliners where the ladies can hang out, relax, share meals or snacks, socialize, watch a show on the screen or play games. The facility is staffed 24/7, features specialized bathrooms, and safety glass throughout.

Ladies Community Building

Country Breeze Farm features a unique multi-use building with a stage that allows for worship services, cafeteria dining, indoor recreational activities, and entertainment events such as plays and musicals. The building also has a fitness center, classrooms for arts and education, administrative offices, a kitchen, reception area, bathrooms, and maintenance and storage facilities.

Recreation & Events

Country Breeze Farm features a campus designed and filled with peaceful places, outdoor lunch & picnic areas, and more. The outdoor recreation areas include places for bocce ball, basketball, volleyball, frisbee, soccer, walking paths, fishing, paddleboats & canoeing, picnic outings, and places for music concerts and festivals. Indoor recreational events and activities are just as plentiful and includes a fitness facility, music, as well as many other activities.

Greenhouses & Aquaponics

Greenhouses and aquaponics on-site allow the ladies to actively work at producing healthy food, raise fish, and grow flowers.

Chicken House & Homing Pigeons

Chickens and homing pigeons provide further animal interaction. Fresh eggs and are available for our ladies to eat and sell.

Outdoor Organic Farming

Healthy food production extends out to the fully functional oudoor organic gardening areas. The ladies and staff enjoy eating the produce they grow, and the satisfaction of selling it at the Country Breeze Market.

Indoor Riding Arena & Barn for Animals

Country Breeze Farms features the opportunity for Therapeutic Horseback Riding. This allows the ladies to bond in a special way with the animals. An outdoor fenced in area is also available for riding horses and keeping goats.

Country Breeze Market

Country Breeze Market provides additional occupational opportunities for our special ladies. It provides additional opportunities for socialization and supervised interaction with customers. The store is geared towards educating and helping others become aware of the special needs community and the importance of Country Breeze Farm. The store also helps raise funds for the ladies.

In addition, Country Breeze Market is a very practical way to teach and promote the importance of sustainable organic farming practices. It provides the local community with natural, fresh, organic and healthy produce.

Environmentally Friendly Campus

Country Breeze Farms is an environmentally friendly campus. It features permeable driving and parking areas with natural filtration and buffers to mitigate storm water runoff and pollution. Buildings are also energy efficient and contain appliances and equipment that reduce the impact on the environment. The facility emphasizes recycling, conservation, and the use of environmentally friendly products.

Programs and Services

Country Breeze Farm encompasses a safe and comfortable environment with programs and services that enables our special ladies to be blessed with a good quality life with purpose, love, and nurturing companionship. Our around the clock all female staff help our special ladies discover and enjoy occupational projects, whether it be caring for the land and animals, cultivating and harvesting organic produce and flowers, helping in the office, or working in other ways on campus. Qualified staff assist the ladies participate in many different activities, such as therapeutic horseback riding, exercising in the fitness facility, crafting, music program, indoor/outdoor recreational activities, excursions into town, speech/occupational therapy, and daily chores.

Continued Growth and Self-Improvement

Being in my fifth decade of life I am able to look back and recognize certain things which really made a positive impact on me. Beyond education, life experiences, and people who played an instrumental role in my life, I have been blessed beyond measure by certain documentaries and books. These resources have either greatly inspired me or better educated me to a point of positive change. I wanted to dedicate this portion of this book to pay it forward in the hope that you may be able to benefit from them as well. Following you will find resources that have truly improved my quality of life. I very much recommend the below resources and hope that you will find value in them, too.

Documentaries

I absolutely love documentaries. I learn so much from them. Following are my favorite documentaries that I highly recommend. Most of these recommended documentaries received a review rating of at least 4.5 stars out of 5 stars.

Food / Nutrition / Health Documentaries (Highly Recommended)

- Cowspiracy
- Eating - 3rd Edition (by Mike Anderson)
- Fat, Sick & Nearly Dead
- Fat, Sick & Nearly Dead 2
- Fed Up
- Food Chains
- Food Choices
- Food Matters
- Food, Inc.
- Forks Over Knives
- Fresh
- GMO OMG
- Hungry for Change
- Killer at Large
- King Corn
- Plant Pure Nation
- Processed People
- Scientists Under Attack- Genetic Engineering in the Magnetic Field of Money
- Sugar Coated
- Supersize Me
- The Beautiful Truth (nutrition for cancer patients)
- The Future of Food
- The Gerson Miracle
- The Kids Menu
- The Weight of a Nation
- Vegucated

Social, Environmental, or Nature Documentaries (Highly Recommended)

- (Dis)Honesty: The Truth About Lies
- A Crude Awakening
- Bag It
- Blue Gold: World Water Wars
- Cowspiracy
- Earthlings (by Shaun Monson)
- End of the Line (by film-maker Rupert Murray)
- Flow: For Love of Water
- Gasland
- God of Wonders
- Happy
- In the Womb (National Geographic)
- Inside Job
- Inside Planet Earth
- Life
- Living on One Dollar
- Minimalism: A Documentary About the Important Things
- Nature's Most Amazing Events
- No Place on Earth
- Planet Earth
- Plastic Paradise: The Great Pacific Garbage Patch
- Poverty, Inc.
- SlingShot
- Tapped
- The College Conspiracy (a documentary on YouTube)
- The Great Rift: Africa's Greatest Story
- The Lee Strobel Film Collection
- The World According to Monsanto
- Vanishing of the bees
- Waste Land directed by Lucy Walker (Arthouse Studio)
- Winter on Fire: Ukraine's Fight for Freedom

Historical Documentaries (Highly Recommended)

- Above and Beyond
- Auschwitz: The Nazis and the 'Final Solution'
- Brothers in War
- Desperate Crossing: Mayflower
- Diaries of the Great War
- Escape from a Nazi Death Camp
- Reader's Digest WWII in the Pacific
- The Civil War
- The First World War (the complete series)
- The Long Way Home
- The Longest Day by 20th Century Fox
- The War
- The World at War
- Treblinka
- World War II - War in the Pacific

Inspirational Films (Highly Recommended)

Below is a list of my favorite inspirational movies.

- Courageous
- Facing the Giants
- Fire Proof
- Flywheel

Classics (Highly Recommended)

Following is my list of favorite classical films. These films received excellent reviews.

- An American in Paris
- Fiddler on the Roof
- It's a Wonderful Life
- Oklahoma!
- Seven Brides for Seven Brothers
- Singin' in the Rain
- The General (this silent movie filmed in 1926 is the funniest movie I have ever seen)
- The Great Locomotive
- The Music Man
- The Sound of Music
- White Christmas

Books & Audio Books (Highly Recommended)

The below books are the best books I have ever read in my life. Coincidently, most all received a customer review rating of 5 stars out of 5 stars.

- 1776 by David McCullough
- 50/50: Secrets I Learned Running 50 Marathons in 50 Days
- 7 Habits of Highly Successful People
- A Thousand-Mile Walk to the Gulf by John Muir
- As A Man Thinketh by James Allen
- Awaken the Giant Within
- Band of Brothers by Stephen E. Ambrose
- Born to Run: A Hidden Tribe, Superathletes, and the Greatest Race the World Has Never Seen
- Brian Tracy (all books by Brian Tracy are excellent)
- Bringing Up Boys by James Dobson
- Bringing Up Girls by James Dobson
- Caffeine Blues
- China Study
- Coming Back Stronger: Unleashing the Hidden Power of Adversity
- Desiring God by John Piper
- Disciplines of a Godly Man by R. Kent Hughes
- Don't Waste Your Life by John Piper

- Driven: How To Succeed In Business And In Life by Robert Herjavec
- Eat and Run: My Unlikely Journey to Ultramarathon Greatness
- Eat to Live: The Amazing Nutrient-Rich Program for Fast and Sustained Weight Loss by Joel Fuhrman
- Evangelism and the Sovereignty of God, J.I. Packer
- Extreme Pursuit by John E. Davis
- From Pride to Humility by Stuart Scott
- God's Wisdom in Proverbs by Phillips
- Good to Great: Why Some Companies Make the Leap and Others Don't
- Grace to You by John MacArthur
- Handwriting of the Famous and Iinfamous by Sheila Lowe
- Happy is the man by Robert V. Ozment
- Happy, Happy, Happy
- Have a New Kid by Friday by Leman
- Healthy Eating, Healthy World: Unleashing the Power of Plant-Based Nutrition, by J. Morris Hicks
- How Successful People Think by John Maxwell
- How to Win Friends & Influence People, Dale Carnegie
- Lincoln the Unknown, Dale Carnegie
- Love Dare
- Love for a Lifetime: Building a Marriage That Will Go the Distance
- Making Men by Chuck Holton
- Men Are from Mars, Women Are from Venus
- No Happy Cows by John Robbins
- Nothing to Envy
- One Minute Manager
- Parenting Collection by James Dobson
- Pursuit of Holiness by Jerry Bridges
- Quiet Strength by Tony Dungy
- Raising a Modern-Day Knight by Robert Lewis (book for Dad's with sons)
- Remember Names by Dale Carnegie
- Respectable Sins by Jerry Bridges (*best book I have ever read in my life*)
- Rich Dad, Poor Dad
- Running Man: A Memoir
- Seeking Allah, finding Jesus
- Shaken: Discovering Your True Identity in the Midst of Life's Storms
- Shepherding a Child's Heart by Tedd Tripp
- Strong Willed Child by James Dobson
- The 10 natural laws of successful time and life management
- The 5 Love Languages: The Secret to Love that Lasts
- The Attributes of God, Arthur W. Pink
- The Autobiography of Benjamin Franklin
- The Backyard Homestead
- The Endurance: Shackleton's Legendary Antarctic Expedition
- The Exemplary Husband by Stuart Scott
- The Gluten Connection: How Gluten Sensitivity May Be Sabotaging Your Health - And What You Can Do to Take Control Now
- The Greatest Miracle in the World by O.G. Mandino
- The Greatest Salesmen in the World
- The Guide to Confident Living, by Norman Vincent Peale
- The Human Body Book (Book & DVD)
- The Marriage You've Always Wanted by Gary Chapman
- The Path Between the Seas: The Creation of the Panama Canal, 1870-1914
- The Personal MBA by Josh Kaufman
- The Power of Positive Thinking by Norman Vincent Peale
- The Power of Positive Thinking, Norman Vincent Peale
- The Psychology of Winning by Dr. Dennis Waitley
- The Quest for Character, John MacArthur
- The Success Principles: How to get from where you are to where you want to be, by Jack Canfield with Janet Switzer.
- The Truth War: Fighting for Certainty in an Age of Deception by John F. MacArthur
- Think and Grow Rich
- Through My Eyes By Tim Tebow
- Ultramarathon Man: Confessions of an All-Night Runner
- Undaunted Courage: Meriwether Lewis, Thomas Jefferson, and the Opening of the American West
- Way of the Master by Ray Comfort
- Whitewash: The Disturbing Truth About Cow's Milk and Your Health

Best of the Best (Highly Recommended)

Although I highly recommend all of the above resources, selecting the best of the best out of each category, following is my 'must see/read' list. It you were only going to try a few on my list I would put the following as the highest priority.

- **Environmental** — the documentaries 'Cowspiracy' and 'Plastic Paradise'.
- **Nutrition/Health/Weight Loss** — 'Food Choices' (documentary) and 'China Study' (book or audiobook).
- **Social** — the documentaries 'Tapped', 'Poverty, Inc.' and ''Flow: for the Love of Water'.

- **Business/Entrepreneurship/Success** — all books by Brain Tracy, and the book 'As A Man Thinketh' by James Allen
- **Spiritual** — the books 'Respectable Sins' and 'Pursuit of Holiness' by Jerry Bridges
- **Nature** — the documentaries 'Planet Earth' and 'The Blue Planet'.
- **Inspirational** — the films noted above in the Inspirational category.
- **Relationships** — 'Fire Proof' (film), 'The 5 Love Languages: the Secret to Love that Last (book), Men are from Mars, Women are from Venus' (book), and 'Love for a Lifetime' book by Dobson).
- **Child Rearing** — 'Parenting Collection' (book by Dobson).

Comment and Feedback

As mentioned previously, I am an avid fan of documentaries and non-fiction books. The above resources have greatly helped me grow as a person and I hope my sharing of these recommendations will help you grow as well. I am still learning and will continue to update this list on the "Farm Your Space" website.

www.FarmYourSpace.com

I welcome your comments, recommendations, and feedback on these and other resources. Thank you.

HALPS
Health And Life Protection Solutions

- **Aquaponics, Organics, Homestead**

- **Nutrition, Weight Loss, Health**

- **Water Testing & Purification**

- **Civil Engineering**

- **Environmental Hazards Solutions**

- **Radon Solutions**

- **EMF Solutions**

- **Mold Solutions**

- **Home Inspections**

The services provided by HALPS will greatly enhance the quality of your life as well as the lives of the people who you love and care about.

Blessed to be able to serve most of east Tennessee.

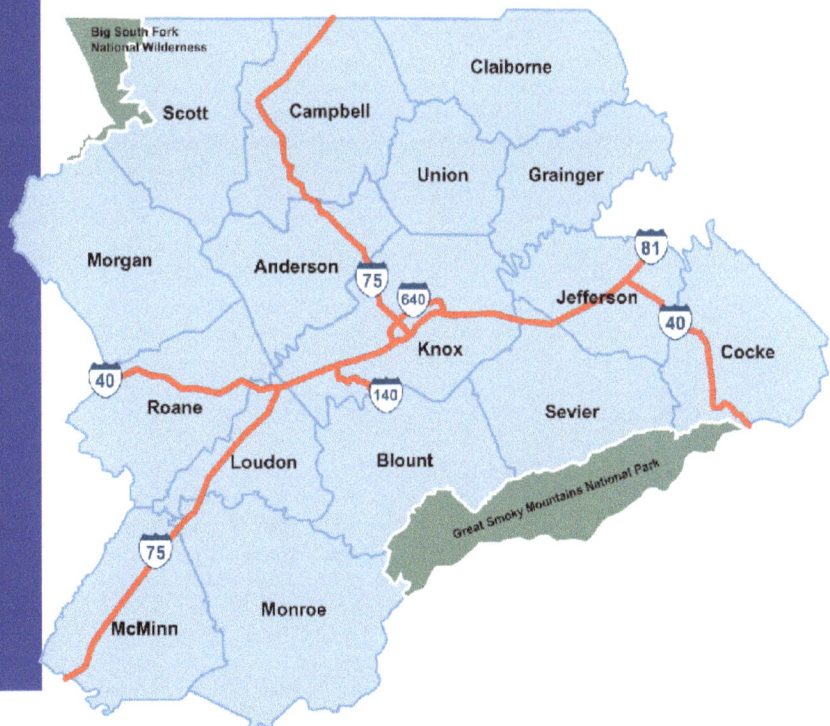

www.HALPS.net

Request for Input

I invite your input, as it will help me grow as a person, and enable me to serve others better. I welcome all contributions, such as your recommended improvements, lessons you've learned, related stories, personal testimony, etc. Please send your feedback, suggestions, contributions, etc. via the website:

www.FarmYourSpace.com

Thank you for your interest in this book. My hope is that it has made a positive contribution to your life. In addition to your feedback, please let me know if and how I can pray for you.

Index